HOW
THE EARTH
WORKS

AN ILLUSTRATED GUIDE TO THE WONDERS OF OUR PLANET

HOW THE EARTH WORKS

AN ILLUSTRATED GUIDE TO THE WONDERS OF OUR PLANET

CHARTWELL
BOOKS

Quarto is the authority on a wide range of topics.

Quarto educates, entertains and enriches the lives of
our readers—enthusiasts and lovers of hands-on living.

www.quartoknows.com

HOW THE EARTH WORKS
This edition published in 2016 by
CHARTWELL BOOKS
an imprint of Book Sales
a division of Quarto Publishing Group USA Inc.
142 West 36th Street, 4th Floor
New York, New York 10018
USA

This edition © 2016 Editorial Sol90, S.L. Barcelona

ISBN-13: 978-0-7858-3439-7

Printed in China

10 9 8 7 6 5 4 3 2 1

Original Idea Nuria Cicero
Editorial Coordination Alberto Hernández
Design Caco Daví, Leandro Jema
Editing Joan Soriano
Layout Carla Cobas

PREFACE

Featuring amazing photography and stunning illustrations in full-color cross sections, *How the Earth Works* reveals all of nature's mysteries. Hundreds of subjects are explored, thousands of facts are revealed, explained, and dissected so that anyone can understand what causes something as simple as fog or as complex as an earthquake. Even the most complicated mysteries—like how and when the planet was formed, the cycle of life, thermoregulation in mammals, and the erosion and transportation of sediments—are clearly illuminated.

Visually-driven and encyclopedic in scope, easy to understand for all readers and even of interest to experts, *How the Earth Works* is organized thematically into eight chapters, and includes topics such as the geology of the planet, volcanoes, earthquakes, ecology, the environment, weather, animals, plants, and natural wonders. You will spend many enjoyable hours discovering all the facts about Earth's inner workings and how some fantastic places like the Amazon, the Himalayas, the Grand Canyon, and Death Valley were formed. Come explore the world, its natural wonders, its hidden secrets, and all its 8.7 million species.

CONTENTS

INTRODUCTION — 8

CHAPTER 1
THE PLANET

The Blue Planet — 14
Movements and Coordinates — 16
The Earth's Magnetism — 18
Eclipses — 20
The Long History of the Earth — 22
Stacked Layers — 24
The Journey of the Plates — 26
Folding in the Earth's Crust — 28
When the Faults Resound — 30
Living Water — 32
Ocean Currents — 34
Jets of Water (Geysers) — 36
Sources of Energy — 38
A Changing Surface — 40
If Stones Could Speak — 42
Metamorphic Processes — 44
Before Rock, Mineral — 46
How to Recognize Minerals — 48
Precious Crystals — 50
How to Identify Rocks — 52
Organic Rocks — 54

CHAPTER 2
VOLCANOES AND EARTHQUAKES

Flaming Furnace — 58
Classification — 60
Flash of Fire — 62
Aftermath of Fury — 64
Latent Danger — 66
Deep Rupture — 68
Elastic Waves — 70
Measuring an Earthquake — 72
Violent Seas — 74

Cause and Effect — 76
Risk Areas — 78

CHAPTER 3
ECOLOGY AND ENVIRONMENT

The Six Kingdoms — 82
The Basis of Life — 84
Ecosystems — 86
Biodiversity — 88
Habitats of the World — 90
Land Biomes — 92
Aquatic Ecosystems — 94
Coral Reefs — 96
The Greenhouse Effect — 98
Climate Change — 100
The Ozone Hole — 102

CHAPTER 4
WEATHER AND CLIMATOLOGY

Global Equilibrium — 106
Climate Zones — 108
Atmosphere Dynamics — 110
Collision — 112
Monsoons — 114
Capricious Forms — 116
The Rain Announces It's Coming — 118
Lost in the Fog — 120
Brief Flash — 122
Lethal Force — 124
Anatomy of a Hurricane — 126

CHAPTER 5
ANIMAL KINGDOM

What is a Mammal? — 130
Constant Heat — 132
Grace and Movement — 134
Extremities — 136

Developed Senses 138
Looks that Kill 140
Herbivores 141
Deep Sleep 144
Record Breath-Holders 146
Nocturnal Flight 148
The Language of Water 150
Beyond Feathers 152
The Senses (Birds) 154
Wings to Fly 156
Fish Anatomy 158
Protective Layer 160
The Art of Swimming 162
Between Land and Water 164
A Skin with Scales 166
Internal Structure (Snakes) 168
Jointless (Mollusks) 170
Colorful Armor (Crustaceans) 172
A Special Family (Arachnids) 174

CHAPTER 6
VEGETAL KINGDOM

Conquest of Land 178
Anatomy of a Tree 180
Feeding on Light 182
Aquatic Plants 184
Seeds, To and Fro 186
Under the Earth 188
Stems: More Than a Support 190
Energy Manufacturers 192
Functional Beauty 194
Pollination 196
Bearing Fruit 198

CHAPTER 7
THE OTHER PLANTS

Colors of Life 202
How Algae Reproduce 204
Terrestrial and Marine Algae 206

Strange Bedfellows 208
Mosses 210
Dispersion of Spores 212
Another World 214
The Diet of Fungi 216
Poison in the Kingdom 218
Pathogens 220
Destroying to Build 222

CHAPTER 8
NATURE WONDERS

The Sahara 226
Death Valley 228
The Himalaya Range 230
Salar de Uyuni 232
The Great Rift Valley 234
The Grand Canyon 236
Iguazu Falls 238
The Antarctic 240
The Amazon 242
Kilimanjaro 244
Cappadocia 246

INDEX
Index 248
Photo credits 256

"

Our population and our use of the finite resources of planet Earth are growing exponentially, along with our technical ability to change the environment for good or ill."

STEPHEN HAWKING

INTRODUCTION

Geologic structures and fossil remains have been used by experts to reconstruct the history of life on our planet. Today it is believed that Earth formed 4.6 billion years ago and that its first living creatures—bacteria—appeared 1 billion years later. From that moment until now, the Earth has registered the emergence, evolution, and extinction of a great number of species.

Earth's first surface was a thin layer covered by volcanoes expelling very light lava from inside. As the lava cooled, it solidified into Earth's initial crust. The gases emitted from these volcanoes formed an early atmosphere, and Earth's first organisms breathed anaerobically. In fact, it is simple one-cell organisms that comprise most of the history of life on Earth. Mainly bacteria, they have survived over 3 billion years. In comparison, the predominance of dinosaurs during the Mesozoic Era (65 million years ago) is a recent invention, and man's presence on Earth is rendered insignificant. Stromatolites, fossils that date to 3.4 billion years ago, are among the first pieces of evidence of early life on the planet. These formations are the product of unicellular algae that lived in shallow water.

Life on Earth was subject to the presence of oxygen which settled on its surface some 2.1 billion years ago. This element enabled the formation of essential compounds such as water and carbon dioxide. Earth's first multi-celled organisms are from the Silurian Period, over 400 million

MAMMALS
Emerged almost 200 million years ago, at the end of the Triassic period. At present there are more than 5,000 species.

MOUNT BATUR
One of the famous actives volcanos in tropical island of Bali, Indonesia. The most recent eruption was in 2000.

MUSHROOMS
These organisms grow in damp, low light environments. Some species are edible and other poisonous.

years ago, when plants invaded sedimentary areas and crustaceans came out of the water.

Amphibians—fish with lungs that allowed them to breathe out of water—were the first terrestrial vertebrates. Subsequently, reptiles (both big and small) dominated the planet, and after their extinction, which marked the end of the Cretaceous Period, a vast number of birds and mammals were able to develop. Grasslands expanded, becoming the predominant biomes and the habitat of humankind's earliest ancestors. The first hominids known to be bi-pedaled (probably *Sahelanthropus Tchadensis*) lived between 6 and 7 million years ago.

The Earth is a living planet in constant transformation. The latest studies estimate that at this moment 8.7 million species live together in our world, among them animals,

plants, fungi, protozoa, and Chromista, with 25 percent residing in our oceans and only 1.4 million currently identified. During millions of years, the continental mass has evolved to a form similar to the one from the Tertiary Period (60 million years ago), at the beginning of the Cenozoic Era when our tallest mountain ranges—the Alps, the Andes, and highest of all, the Himalaya—resulted from the collision of the Indian plate with the Eurasian plate.

Today we are evolved enough to monitor the "heartbeat" of our ever-changing planet, in the form of volcanic activity and earthquakes, in addition to the weather, one of the main agents responsible for changes in the earth's surface. But what cannot be ignored are our very own actions; the most voracious consumers of the Earth are greatly responsible for the phenomenon that we know as climate change.

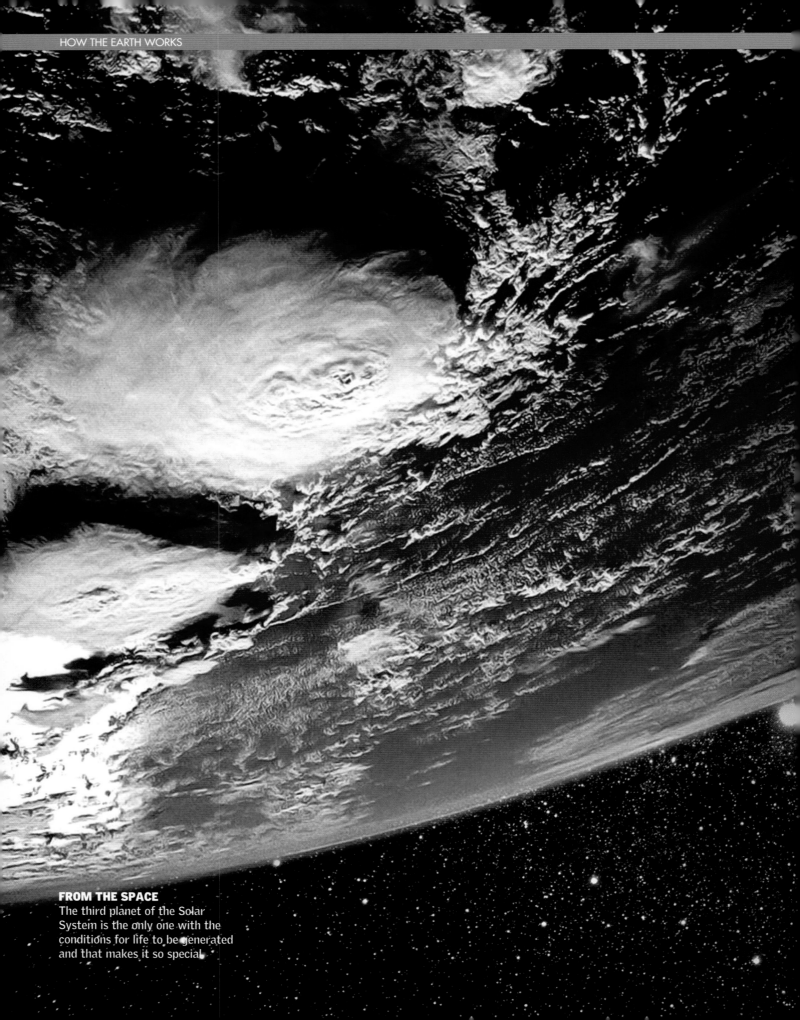

FROM THE SPACE
The third planet of the Solar
System is the only one with the
conditions for life to be generated
and that makes it so special

CHAPTER 1

THE
PLANET

The Earth was formed 4,600 million years ago from a cloud of dust and gas. At first, it was a fiery mass and in constant turmoil, but over time, the Earth began to cool down and the atmosphere got clean as the rain fell and the oceans were created.

The Blue Planet

The Earth is known as the blue planet because of the color of the oceans that cover two thirds of its surface. This planet, the third planet from the Sun, is the only one where the right conditions exist to sustain life, something that makes the Earth special. It has liquid water in abundance, a mild temperature, and an atmosphere that protects it from objects that fall from outer space. The atmosphere also filters solar radiation thanks to its ozone layer. Slightly flattened at its poles and wider at its equator, the Earth takes 24 hours to revolve once on its axis.

The Phenomenon of Life

Water, in liquid form, makes it possible for life to exist on the Earth, the only planet where temperatures vary from 32° F to 212° F (0° C to 100° C), allowing water to exist as a liquid. The Earth's average distance from the Sun, along with certain other factors, allowed life to develop 3.8 billion years ago.

-76° F
(-60° C)

ONLY ICE
Mars is so far from the Sun that all its water is frozen.

32° to 212° F
(0° to 100° C)

3 STATES
On the Earth, water is found in all three of its possible states.

Above 212° F
(100° C)

ONLY STEAM
On Mercury or Venus, which are very close to the Sun, water would evaporate.

EARTH MOVEMENTS

The Earth moves, orbiting the Sun and rotating on its own axis.

SUN

93,500,000 miles
(149,503,000 km)

ROTATION: The Earth revolves on its axis in 23 hours and 56 minutes.

REVOLUTION: It takes the Earth 365 days, 5 hours, and 57 minutes to travel once around the Sun.

The Moon, our only natural satellite, is four times smaller than the Earth and takes 27.32 days to orbit the Earth.

SOUTH POLE

AXIS
INCLINATION

ROTATION
AXIS

NORTH
POLE

CHARACTERISTICS

CONVENTIONAL
PLANET
SYMBOL

Density	3.2 ounces per cubic inch (5.52 g/cu cm)
Average temperature	59° F (15° C)

*In both cases, Earth = 1

ESSENTIAL DATA

Average distance to the Sun	93 million miles (150 million km)
Revolution around the Sun (Earth year)	365.25 days
Diameter at the equator	7,930 miles (12,756 km)
Orbiting speed	17 miles per second (27.79 km/s.)
Mass*	1
Gravity*	1

AXIS INCLINATION

23.5°

One rotation
lasts 23.56
hours.

70%

of the Earth's surface
is water. From space,
the planet looks blue.

23.5°

This is the inclination of the Earth's
axis from the vertical. As the Earth
orbits the Sun, different regions
gradually receive more or less
sunlight, causing the four seasons.

1 **EVAPORATION**
Because of the Sun's energy, the water
evaporates, entering the atmosphere from
oceans and, to a lesser extent, from lakes,
rivers, and other sources on the continents.

2 **CONDENSATION**
The Earth's winds transport moisture-
laden air until weather conditions cause
the water vapor to condense into clouds
and eventually fall to the ground as rain or
other forms of precipitation.

3 **PRECIPITATION**
The atmosphere loses water through
condensation. Gravity causes rain, snow,
and hail. Dew and frost directly alter the
state of the surface they cover.

The Biosphere

Only a small part of the Earth is inhabited by living things: the surface,
the oceans, the first 5 miles (8 km) of the atmosphere above the
ground, and the area beneath the ground as far as plant roots reach.
The biosphere makes up this tiny portion of the planet. Studying the
biosphere helps to reveal the patterns by which different forms of life
became established and the parameters that affect the distribution of
species and ecosystems.

Movements and Coordinates

Yes, it moves. The Earth rotates on its axis while simultaneously orbiting the Sun. The natural phenomena of night and day, seasons, and years are caused by these movements. To track the passage of time, calendars, clocks, and time zones were invented. Time zones are divided by meridians and assigned a reference hour according to their location. When traveling east, an hour is added with each time zone. An hour is subtracted during west-bound travel.

The Earth's Movements

Night and day, summer and winter, new year and old year result from the Earth's various movements during its orbit of the Sun. The most important motions are the Earth's daily rotation from west to east on its own axis and its revolution around the Sun. (The Earth follows an elliptical orbit that has the Sun at one of the foci of the ellipse, so the distance to the Sun varies slightly over the course of a year.)

23.5°

ROTATION
1 DAY
The Earth revolves once on its axis in 23 hours and 56 minutes. We see this as day and night.

REVOLUTION
1 YEAR
The Earth's orbit around the Sun lasts 365 days, 5 hours, and 57 minutes.

9°

NUTATION
18.6 YEARS
is a sort of nod made by the Earth, causing the displacement of the geographic poles by nine arc seconds.

47°

PRECESSION
25,800 YEARS
A slow turning of the direction of the Earth's axis (similar to that of a top), caused by the Earth's nonspherical shape and the gravitational forces of

EQUINOX AND SOLSTICE

Every year, around June 21, the Northern Hemisphere reaches its maximum inclination toward the Sun (a phenomenon referred to as the summer solstice in the Northern Hemisphere and the winter solstice in the Southern Hemisphere). The North Pole receives sunlight all day, while the South Pole is covered in darkness. Between one solstice and another the equinoxes appear, which is when the axis of the Earth points toward the Sun and the periods of daylight and darkness are the same all over our planet.

June 20 or 21
SUMMER SOLSTICE IN THE NORTHERN HEMISPHERE AND WINTER SOLSTICE IN THE SOUTHERN HEMISPHERE
Solstices exist because of the tilt of the Earth's axis. The length of the day and the height of the Sun in the sky are greatest in summer and least in winter.

SUN

March 20 or 21
SPRING EQUINOX IN THE NORTHERN HEMISPHERE AND AUTUMN EQUINOX IN THE SOUTHERN HEMISPHERE
The Sun passes directly above the equator, and day and night have the same length.

September 21 or 22
AUTUMN EQUINOX IN THE NORTHERN HEMISPHERE AND SPRING EQUINOX IN THE SOUTHERN HEMISPHERE
The Sun passes directly above the equator, and day and night have the same length.

MEASUREMENT OF TIME
Months and days are charted by calendars and clocks, but the measurement of these units of time is neither a cultural nor an arbitrary construct. Instead, it is derived from the

PERIHELION
The point where the orbiting Earth most closely approaches the Sun

23.5º
Tilt of the earth's axis

93
MILLION MILES
(149 MILLION KM)

THE EARTH'S ORBIT
About 365 days.

1 day **THE DAYS**
Period of time it takes the Earth to rotate on its axis.

About 30 days **THE MONTHS**
Each period of time, between 28 and 31 days, into which a year is divided.

December 21 or 22

WINTER SOLSTICE IN THE NORTHERN HEMISPHERE AND SUMMER SOLSTICE IN THE SOUTHERN HEMISPHERE
Solstices exist because of the tilt of the Earth's axis. The length of the day and the height of the Sun in the sky are greatest in summer and least in winter.

APHELION
The point in the Earth's orbit where it is farthest from the Sun (94 million miles [152 million km]). This occurs at the beginning of July.

Geographic Coordinates

Thanks to the grid formed by the lines of latitude and longitude, the position of any object on the Earth's surface can be easily located by using the intersection of the Earth's equator and the Greenwich meridian (longitude 0°) as a reference point. This intersection marks the midpoint between the Earth's poles.

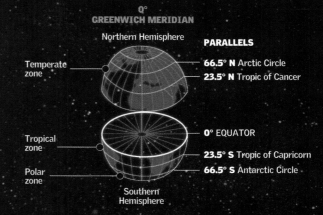

0°
GREENWICH MERIDIAN

Northern Hemisphere

PARALLELS

Temperate zone

66.5° N Arctic Circle
23.5° N Tropic of Cancer

Tropical zone

0° EQUATOR

Polar zone

23.5° S Tropic of Capricorn
66.5° S Antarctic Circle

Southern Hemisphere

Time Zones

The Earth is divided into 24 areas, or time zones, each one of which corresponds to an hour assigned according to the Coordinated Universal Time (UTC), using the Greenwich, England, meridian as the base meridian. One hour is added when crossing the meridian in an easterly direction, and one hour is subtracted when traveling west.

JET LAG

The human body's biological clock responds to the rhythms of light and dark based on the passage of night and day. Long air flights east or west interrupt and disorient the body's clock, causing a disorder known as jet lag. It can cause fatigue, irritability, nausea, headaches, and difficulty sleeping at night.

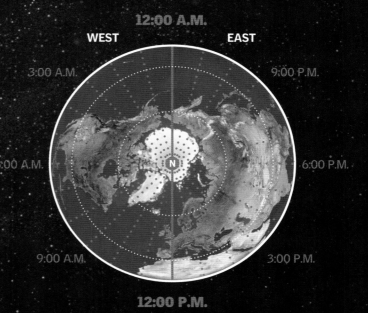

12:00 A.M.

WEST   EAST

3:00 A.M.   9:00 P.M.

6:00 A.M.   6:00 P.M.

N

9:00 A.M.   3:00 P.M.

12:00 P.M.

12:00 A.M.
Departure time

Northern Hemisphere

12:00 P.M.
Arrival time

12:00 15:00 18:00 21:00 0:00 3:00 6:00 9:00

The Earth's Magnetism

The Earth behaves like a giant bar magnet and has a magnetic field with two poles. It is likely that the Earth's magnetism results from the motion of the iron and nickel in its electroconductive core. Another probable origin of the Earth's magnetism lies in the convection currents caused by the heat of the core. The Earth's magnetic field has varied over the course of time. During the last five million years, more than 20 reversals have taken place. The most recent one occurred 700,000 years ago.

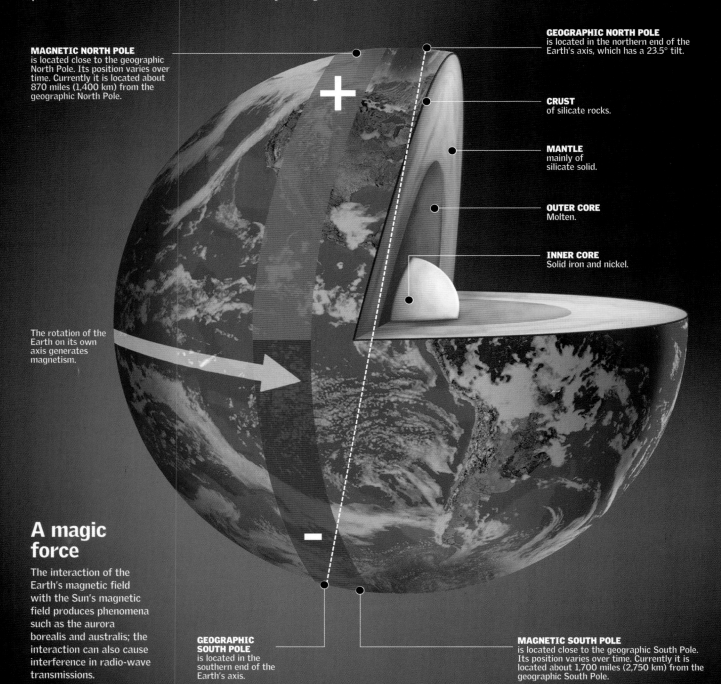

MAGNETIC NORTH POLE
is located close to the geographic North Pole. Its position varies over time. Currently it is located about 870 miles (1,400 km) from the geographic North Pole.

GEOGRAPHIC NORTH POLE
is located in the northern end of the Earth's axis, which has a 23.5° tilt.

CRUST
of silicate rocks.

MANTLE
mainly of silicate solid.

OUTER CORE
Molten.

INNER CORE
Solid iron and nickel.

The rotation of the Earth on its own axis generates magnetism.

A magic force

The interaction of the Earth's magnetic field with the Sun's magnetic field produces phenomena such as the aurora borealis and australis; the interaction can also cause interference in radio-wave transmissions.

GEOGRAPHIC SOUTH POLE
is located in the southern end of the Earth's axis.

MAGNETIC SOUTH POLE
is located close to the geographic South Pole. Its position varies over time. Currently it is located about 1,700 miles (2,750 km) from the geographic South Pole.

The atmosphere
reaches 560 miles
(900 km).

MAGNETOSPHERE

The invisible lines of force that
form around the Earth. It has
an ovoid shape and extends
37,000 miles (60,000 km)
from the Earth. Among other
things, it protects the Earth
from harmful particles radiated
by the Sun.

Solar wind
with charged
atomic
particles

The deformation of the
magnetosphere is caused
by the action of electrically
charged particles
streaming from the Sun.

The Van Allen belts
are bands of ionized
atomic particles.

PLANETARY AND SOLAR MAGNETISM

The planets in the solar system have various magnetic fields with
varying characteristics. The four giant planets possess stronger
magnetic fields than the Earth.

| NEPTUNE | URANUS | SATURN | JUPITER | MARS | EARTH | VENUS | MERCURY | SUN |

It is believed
that in the past
its magnetic
field was
stronger.

It is the only
planet in the
solar system
that does
not have a
magnetic field.

It has
a weak
magnetic
field.

Gravity

This is the name given to the mutual attraction of two objects with mass. It is one of the four fundamental forces
observed in nature. The effect of gravity on a body tends to be associated, in common language, with the concept
of weight. Gravity is responsible for large-scale movements throughout the universe; it causes, for example, the
planets in the solar system to orbit the Sun.

The gases that
flow from the Sun's
corona produce
a magnetic field
around it.

WEIGHT

Weight is the force of
the gravity that acts
on a body.

24 pounds (11 kg)

ON THE MOON
The Moon has less mass
than the Earth and, as a
result, less gravity.

154 pounds (70 kg)

ON EARTH
The object is drawn toward
the Earth's center.

390 pounds (177 kg)

ON JUPITER
Jupiter has 300 times more
mass than the Earth and
therefore more gravity.

Eclipses

Typically four times a year, during the full or new moon, the centers of the Moon, the Sun, and the Earth become aligned, causing one of the most marvelous celestial phenomena: an eclipse. At these times, the Moon either passes in front of the Sun or passes through the Earth's shadow. The Sun —even during an eclipse—is not safe to look at directly, since it can cause irreparable damage to the eyes, such as burns on the retina. Special high-quality filters or indirect viewing by projecting the Sun's image on a sheet of paper are some of the ways in which this celestial wonder can be watched. Solar eclipses provide, in addition, a good opportunity for astronomers to conduct scientific research.

TOTAL LUNAR ECLIPSE, SEEN FROM THE EARTH
The orange color comes from sunlight that has been refracted and colored by the Earth's atmosphere.

ANNULAR ECLIPSE OF THE SUN, SEEN FROM THE EARTH

Solar Eclipse

Solar eclipses occur when the Moon passes directly between the Sun and the Earth, casting a shadow along a path on the Earth's surface. The central cone of the shadow is called the umbra, and the area of partial shadow around it is called the penumbra. Viewers in the regions where the umbra falls on the Earth's surface see the Moon's disk completely obscure the Sun—a total solar eclipse. Those watching from the surrounding areas that are located in the penumbra see the Moon's disk cover only part of the Sun—a partial solar eclipse.

ALIGNMENT

Sun   Moon Earth

During a solar eclipse, astronomers take advantage of the blocked view of the Sun in order to use devices designed to study the Sun's atmosphere.

TYPES OF ECLIPSES

TOTAL
The Moon is between the Sun and the Earth and creates a cone-shaped shadow.

ANNULAR
The Sun appears larger than the Moon, and it remains visible around it.

PARTIAL
The Moon does not cover the Sun completely, so the Sun appears as a crescent.

SUN'S APPARENT SIZE

400 times larger than the Moon

SUNLIGHT

DISTANCE FROM THE SUN TO THE EARTH

400 times greater than the distance from the Earth to the Moon.

Lunar Eclipse

When the Earth passes directly between the full Moon and the Sun, a lunar eclipse (which could be total, partial, or penumbral) occurs. Without the Earth's atmosphere, during each lunar eclipse, the Moon would become completely invisible (something that never happens). The totally eclipsed Moon's characteristic reddish color is caused by light refracted by the Earth's atmosphere. During a partial eclipse, on the other hand, part of the Moon falls in the shadow cone, while the rest is in the penumbra, the outermost, palest part. It is not dangerous to look at a lunar eclipse directly.

ALIGNMENT

Sun   Earth   Moon

During an eclipse, the Moon is not completely black but appears reddish.

TYPES OF ECLIPSES

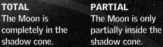

TOTAL
The Moon is completely in the shadow cone.

PARTIAL
The Moon is only partially inside the shadow cone.

PENUMBRAL
The Moon is in the penumbral cone.

Shadow cone

Lunar orbit

FULL MOON
TOTAL ECLIPSE

PARTIAL ECLIPSE

PENUMBRAL ECLIPSE

Penumbra cone

Shadow cone

EARTH

NEW MOON
TOTAL ECLIPSE

Earth orbit

THE ECLIPSE CYCLE

Eclipses repeat every 223 lunations—18 years and 11 days. These are called Saros periods.

ECLIPSES IN A YEAR			ECLIPSES IN A SAROS		
2	7	4	41	29	70
Minimum	Maximum	Average	of the Sun	of the Moon	Total

OBSERVATION FROM EARTH

A black, polymer filter, with an optical density of 5.0, produces a clear orange image of the Sun.

Prevents retinal burns

SOLAR ECLIPSES
are different for each local observer.

MAXIMUM DURATION
8 minutes

LUNAR ECLIPSES
are the same for all observers.

MAXIMUM DURATION
100 minutes

ECLIPSES IN 2006 AND BEYOND

OF THE SUN:
- 3/29 Total
- 9/22 Total
- 3/19 Partial
- 9/11 Partial
- 2/07 Total
- 1/26 Total
- 7/22 Total
- 1/15 Total
- 7/11 Total
- 1/4 Partial
- 11/25 Partial
- 5/20 Annular
- 11/13 Annular
- 5/10 Annular
- 11/3 Hybrid
- 4/29 Annular
- 10/23 Partial
- 3/20 Total
- 9/13 Partial

Years: 2006 | 2007 | 2008 | 2009 | 2010 | 2011 | 2012 | 2013 | 2014 | 2015 | 2016

OF THE MOON:
- 3/14 Partial
- 9/07 Partial
- 3/03 Total
- 8/28 Total
- 2/21 Total
- 8/16 Partial
- 2/9 Partial
- 7/7 Partial
- 6/26 Partial
- 12/21 Total
- 6/15 Total
- 12/10 Total
- 6/4 Partial
- 12/28 Partial
- 4/25 Partial
- 10/18 Partial
- 4/15 Total
- 10/08 Total
- 4/4 Total
- 9/28 Total

The Long History of the Earth

The nebular hypothesis developed by astronomers suggests that the Earth was formed in the same way and at the same time as the rest of the planets and the Sun. It all began with an immense cloud of helium and hydrogen and a small portion of heavier materials 4.6 billion years ago. Earth emerged from one of these "small" revolving clouds, where the particles constantly collided with one another, producing very high temperatures. Later, a series of processes took place that gave the planet its present shape.

From Chaos to Today's Earth

Earth was formed 4.6 billion years ago. In the beginning it was a body of incandescent rock in the solar system. The first clear signs of life appeared in the oceans 3.6 billion years ago, and since then life has expanded and diversified. The changes have been unceasing, and, according to experts, there will be many more changes in the future.

4.5
BILLION YEARS AGO
COOLING
The first crust formed as it was exposed to space and cooled. Earth's layers became differentiated by their density.

4.6 **BILLION YEARS AGO**

FORMATION
The accumulation of matter into solid bodies, a process called accretion, ended, and the Earth stopped increasing in volume.

60
MILLION YEARS AGO
FOLDING IN THE TERTIARY PERIOD
The folding began that would produce the highest mountains that we now have (the Alps, the Andes, and the Himalayas) and that continues to generate earthquakes even today.

540
MILLION YEARS AGO
PALEOZOIC ERA FRAGMENTATION
The great landmass formed that would later fragment to provide the origin of the continents we have today. The oceans reached their greatest rate of expansion.

1.0
BILLION YEARS AGO
SUPERCONTINENTS
Rodinia, the first supercontinent, formed, but it completely disappeared about 650 million years ago.

4
BILLION YEARS AGO
METEORITE COLLISION

Meteorite collisions, at a rate 150 times as great as that of today, evaporated the primitive ocean and resulted in the rise of all known forms of life.

The oldest rocks appeared.

3.8
BILLION YEARS AGO
ARCHEAN EON
STABILIZATION

The processes that formed the atmosphere, the oceans, and protolife intensified. At the same time, the crust stabilized, and the first plates of Earth's crust appeared. Because of their weight, they sank into Earth's mantle, making way for new plates, a process that continues today.

When the first crust cooled, intense volcanic activity freed gases from the interior of the planet, and those gases formed the atmosphere and the oceans.

THE AGE OF THE SUPER VOLCANOES

Indications of komatite, a type of igneous rock that no longer exists.

1.8
BILLION YEARS AGO
PROTEROZOIC EON
CONTINENTS

The first continents, made of light rocks, appeared. In Laurentia (now North America) and in the Baltic, there are large rocky areas that date back to that time.

2.2
BILLION YEARS AGO
WARMING

Earth warmed again, and the glaciers retreated, giving way to the oceans, in which new organisms would be born. The ozone layer began to form.

2.3
BILLION YEARS AGO
"SNOWBALL" EARTH

Hypothesis of a first, great glaciation.

Stacked Layers

Every 110 feet (33 m) below the Earth's surface, the temperature increases by 1.8 degrees Fahrenheit (1 degree Celsius). To reach the Earth's center —which, in spite of temperatures above 12,000° F (6,700° C), is assumed to be solid because of the enormous pressure exerted on it— a person would have to burrow through four well-defined layers. The gases that cover the Earth's surface are also divided into layers with different compositions. Forces act on the Earth's crust from above and below to sculpt and permanently alter it.

Earth's crust

Earth's crust is its solid outer layer, with a thickness of 3 to 9 miles (4 to 15 km) under the oceans and up to 44 miles (70 km) under mountain ranges. Volcanoes on land and volcanic activity in the mid-ocean ridges generate new rock, which becomes part of the crust. The rocks at the bottom of the crust tend to melt back into the rocky mantle.

KEY ● Sedimentary Rock ● Igneous Rock ● Metamorphic Rock

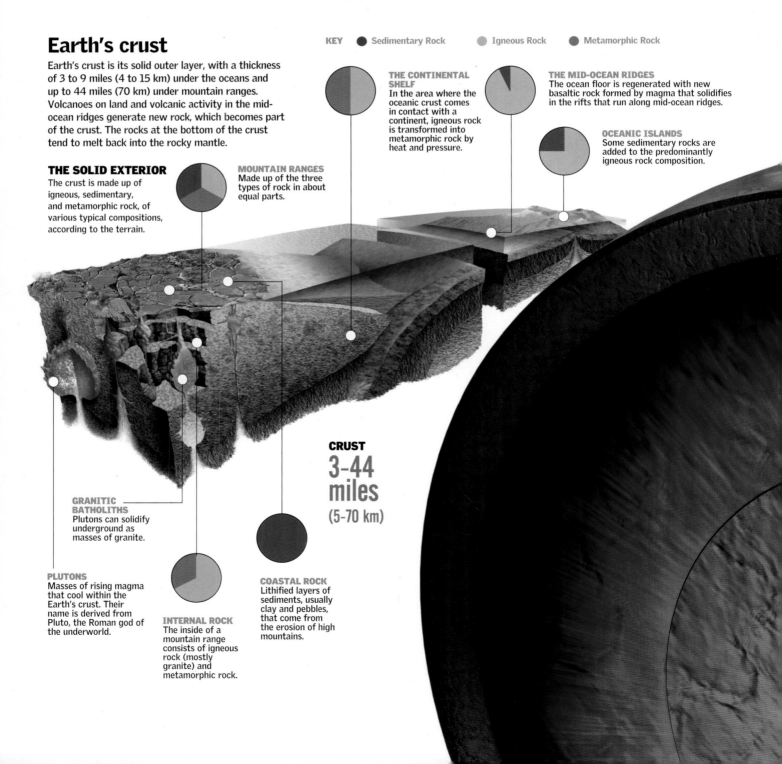

THE CONTINENTAL SHELF
In the area where the oceanic crust comes in contact with a continent, igneous rock is transformed into metamorphic rock by heat and pressure.

THE MID-OCEAN RIDGES
The ocean floor is regenerated with new basaltic rock formed by magma that solidifies in the rifts that run along mid-ocean ridges.

OCEANIC ISLANDS
Some sedimentary rocks are added to the predominantly igneous rock composition.

THE SOLID EXTERIOR
The crust is made up of igneous, sedimentary, and metamorphic rock, of various typical compositions, according to the terrain.

MOUNTAIN RANGES
Made up of the three types of rock in about equal parts.

CRUST
3-44 miles
(5-70 km)

GRANITIC BATHOLITHS
Plutons can solidify underground as masses of granite.

PLUTONS
Masses of rising magma that cool within the Earth's crust. Their name is derived from Pluto, the Roman god of the underworld.

INTERNAL ROCK
The inside of a mountain range consists of igneous rock (mostly granite) and metamorphic rock.

COASTAL ROCK
Lithified layers of sediments, usually clay and pebbles, that come from the erosion of high mountains.

The Gaseous Envelope

The air and most of the weather events that affect our lives occur only in the lower layer of the Earth's atmosphere. This relatively thin layer, called the troposphere, is up to 11 miles (18 km) thick at the equator but only 5 miles (8 km) thick at the poles. Each layer of the atmosphere has a distinct composition.

Less than
11 miles
(18 km)
TROPOSPHERE
Contains 75 percent of the gas and almost all of the water vapor in the atmosphere.

Less than
31 miles
(50 km)
STRATOSPHERE
Very dry; water vapor freezes and falls out of this layer, which contains the ozone layer.

Less than
50 miles
(80 km)
MESOSPHERE
The temperature is -130° F (-90° C), but it increases gradually above this layer.

Less than
280 miles
(450 km)
THERMOSPHERE
Very low density. Below 155 miles (250 km) it is made up mostly of nitrogen; above that level it is mostly oxygen.

Greater than
300 miles
(480 km)
EXOSPHERE
No fixed outer limit. It contains lighter gases such as hydrogen and helium, mostly ionized.

UPPER MANTLE
440 miles
(710 km)

LOWER MANTLE
1,400 miles
(2,250 km)
Composition similar to that of the crust, but in a liquid state and under great pressure, between 1,830° and 8,130° F (1,000° and 4,500° C).

OUTER CORE
1,475 miles
(2,375 km)
Composed mainly of molten iron and nickel among other metals at temperatures above 8,500° F (4,700° C).

INNER CORE
683 miles
(1,100 km)
The inner core behaves as a solid because it is under enormous pressure.

LITHOSPHERE ———— 62 miles
Includes the solid outer part of the upper mantle, as well as the crust. (100 km)

ASTHENOSPHERE ———— 300 miles
Underneath is the asthenosphere, made up of partially molten rock. (480 km)

The Journey of the Plates

When geophysicist Alfred Wegener suggested in 1910 that the continents were moving, the idea seemed fantastic. There was no way to explain the idea. Only a half-century later, plate tectonic theory was able to offer an explanation of the phenomenon. Volcanic activity on the ocean floor, convection currents, and the melting of rock in the mantle power the continental drift that is still molding the planet's surface today.

Continental Drift

The first ideas on continental drift proposed that the continents floated on the ocean. That idea proved inaccurate. The seven tectonic plates contain portions of ocean beds and continents. They drift atop the molten mantle like sections of a giant shell. Depending on the direction in which they move, their boundaries can converge (when they tend to come together), diverge (when they tend to separate), or slide horizontally past each other (along a transform fault).

The Hidden Motor

Convection currents of molten rock propel the crust. Rising magma forms new sections of crust at divergent boundaries. At convergent boundaries, the crust melts into the mantle. Thus, the tectonic plates act like a conveyor belt on which the continents travel.

... 250 million years ago
The landmass today's continents come from was a single block (Pangea) surrounded by the ocean.

PANGEA

... 180 million years ago
The North American Plate has separated, as has the Antarctic Plate. The supercontinent Gondwana (South America and Africa) has started to divide and form the South Atlantic. India is separating from Africa.

LAURASIA

GONDWANA

ANTARCTICA

2 inches (5 cm)

Typical distance the plates travel in a year.

Indo-Australian Plate

CONVERGENT BOUNDARY
When two plates collide, one sinks below the other, forming a subduction zone. This causes folding in the crust and volcanic activity.

Tongan Trench

Eastern Pacific Ridge

Nazca Plate

Peru-Chile Trench

CONVECTION CURRENTS
The hottest molten rock rises; once it rises, it cools and sinks again. This process causes continuous currents in the mantle.

OUTWARD MOVEMENT
The action of the magma causes the tectonic plate to move toward a subduction zone at its far end.

... 100 million years ago

The Atlantic Ocean has formed. India is headed toward Asia, and when the two masses collide, the Himalayas will rise. Australia is separating from Antarctica.

... 60 million years ago

The continents are near their current location. India is beginning to collide with Asia. The Mediterranean is opening, and the folding is already taking place that will give rise to the highest mountain ranges of today.

250 MILLION YEARS

The number of years it will take for the continents to drift together again.

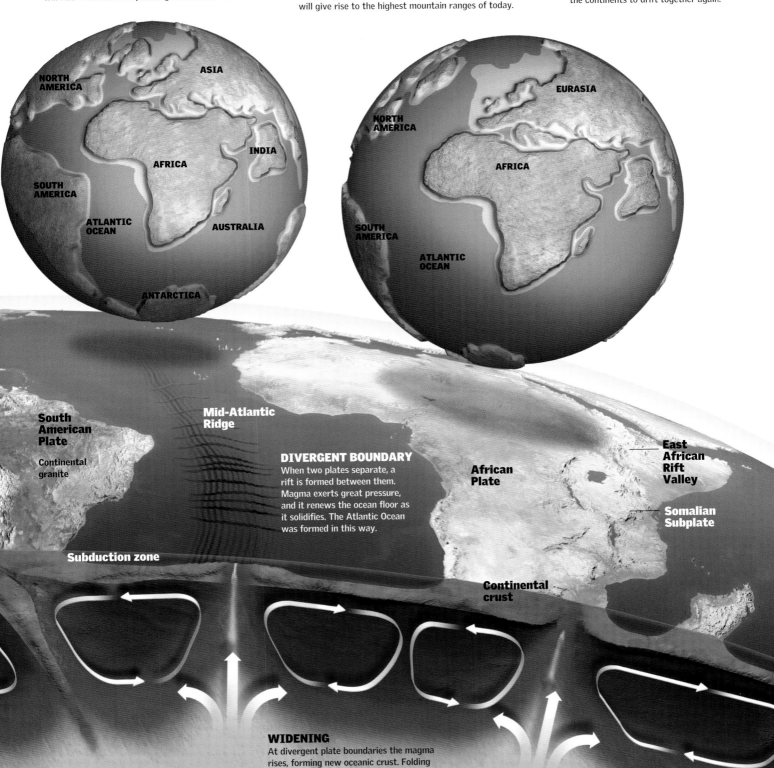

NORTH AMERICA
ASIA
INDIA
AFRICA
SOUTH AMERICA
ATLANTIC OCEAN
AUSTRALIA
ANTARCTICA

EURASIA
NORTH AMERICA
AFRICA
SOUTH AMERICA
ATLANTIC OCEAN

South American Plate

Continental granite

Mid-Atlantic Ridge

DIVERGENT BOUNDARY

When two plates separate, a rift is formed between them. Magma exerts great pressure, and it renews the ocean floor as it solidifies. The Atlantic Ocean was formed in this way.

African Plate

East African Rift Valley

Somalian Subplate

Subduction zone

Continental crust

WIDENING

At divergent plate boundaries the magma rises, forming new oceanic crust. Folding occurs where plates converge.

Folding in the Earth's Crust

The movement of tectonic plates causes distortions and breaks in the Earth's crust, especially in convergent plate boundaries. Over millions of years, these distortions produce larger features called folds, which become mountain ranges. Certain characteristic types of terrain give clues about the great folding processes in Earth's geological history.

Distortions of the Crust

The crust is composed of layers of solid rock. Tectonic forces, resulting from the differences in speed and direction between plates, make these layers stretch elastically, flow, or break. Mountains are formed in processes requiring millions of years. Then external forces, such as erosion from wind, ice, and water, come into play. If slippage releases rock from the pressure that is deforming it elastically, the rock tends to return to its former state and can cause earthquakes.

1 A portion of the crust subjected to a sustained horizontal tectonic force is met by resistance, and the rock layers become deformed.

2 The outer rock layers, which are often more rigid, fracture and form a fault. If one rock boundary slips underneath another, a thrust fault is formed.

3 The composition of rock layers shows the origin of the folding, despite the effects of erosion.

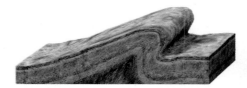

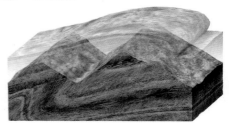

The Three Greatest Folding Events

The Earth's geological history has included three major mountain-building processes, called "orogenies." The mountains created during the first two orogenies (the Caledonian and the Hercynian) are much lower today because they have undergone millions of years of erosion.

MATERIALS Mostly granite, slate, amphibolite, gneiss, quartzite, and schist.

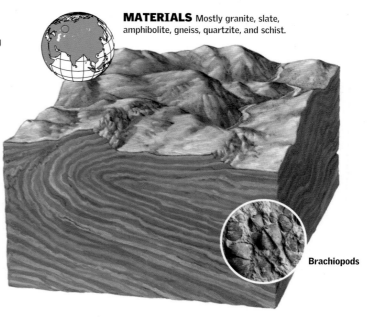

Brachiopods

MATERIALS Mudstone, slate, and sandstone, in lithified layers.

Trilobites

430 Million Years
CALEDONIAN OROGENY
Formed the Caledonian range. Remnants can be seen in Scotland, the Scandinavian Peninsula, and Canada (which all collided at that time).

300 Million Years
HERCYNIAN OROGENY
Took place between the late Devonic and the early Permian periods. It was more important than the Caledonian orogeny. It shaped central and western Europe and produced large veins of iron ore and coal. This orogeny gave rise to the Ural Mountains, the Appalachian range in North America, part of the Andes, and Tasmania.

Formation of the Himalayas

The highest mountains on Earth were formed following the collision of India and Eurasia. The Indian Plate is sliding horizontally underneath the Asiatic Plate. A sedimentary block trapped between the plates is cutting the upper part of the Asiatic Plate into segments that are piling on top of each other. This folding process gave rise to the Himalayan range, which includes the highest mountain on the planet, Mount Everest (29,035 feet [8,850 m]). This deeply fractured section of the old plate is called an accretion prism. At that time, the Asian landmass bent, and the plate doubled in thickness, forming the Tibetan Plateau.

SOUTHEAST ASIA

India today

10 MILLION YEARS AGO

20 MILLION YEARS AGO

30 MILLION YEARS AGO

Ammonites

MATERIALS
High proportions of sediment in Nepal, batholiths in the Asiatic Plate, and intrusions of new granite: iron, tin, and tungsten.

60 Million Years

ALPINE OROGENY
Began in the Cenozoic Era and continues today. This orogeny raised the entire system of mountain ranges that includes the Pyrenees, the Alps, the Caucasus, and even the Himalayas. It also gave the American Rockies and the Andes Mountains their current shape.

A COLLISION OF CONTINENTS

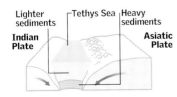

Lighter sediments — Tethys Sea — Heavy sediments

Indian Plate   **Asiatic Plate**

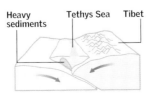

Heavy sediments — Tethys Sea — Tibet

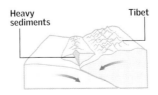

Heavy sediments — Tibet

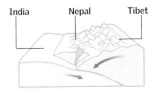

India — Nepal — Tibet

60 MILLION YEARS AGO
The Tethys Sea gives way as the plates approach. Layers of sediment begin to rise.

40 MILLION YEARS AGO
As the two plates approach each other, a subduction zone begins to form.

20 MILLION YEARS AGO
The Tibetan Plateau is pushed up by pressure from settling layers of sediment.

THE HIMALAYAS TODAY
The movement of the plates continues to fold the crust, and the land of Nepal is slowly disappearing.

When the Faults Resound

Faults are small breaks that are produced along the Earth's crust. Many, such as the San Andreas fault, which runs through the state of California, can be seen readily. Others, however, are hidden within the crust. When a fault fractures suddenly, an earthquake results. Sometimes fault lines can allow magma from lower layers to break through to the surface at certain points, forming a volcano.

Relative Movement Along Fault Lines

Fault borders do not usually form straight lines or right angles; their direction along the surface changes. The angle of vertical inclination is called "dip." The classification of a fault depends on how the fault was formed and on the relative movement of the two plates that form it. When tectonic forces compress the crust horizontally, a break causes one section of the ground to push above the other. In contrast, when the two sides of the fault are under tension (pulled apart), one side of the fault will slip down the slope formed by the other side of the fault.

350 miles
(566 km)

The distance that the opposite sides of the fault have slipped past each other, throughout their history.

1

NORMAL FAULT
This fault is the product of horizontal tension. The movement is mostly vertical, with an overlying block (the hanging wall) moving downward relative to an underlying block (the footwall). The fault plane typically has an angle of 60 degrees from the horizontal.

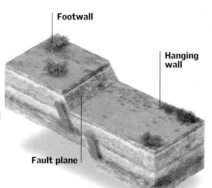

Footwall

Hanging wall

Fault plane

Rodgers Creek

Concord-Green Valley

Mt. Diablo

OAKLAND

Greenville

SAN FRANCISCO

Hayward

Calaveras

San Gregorio

PACIFIC OCEAN

Footwall

Hanging wall

2

REVERSE FAULT
This fault is caused by a horizontal force that compresses the ground. A fracture causes one portion of the crust (the hanging wall) to slide over the other (the footwall). Thrust faults (see pages 28-29), are a common form of reverse fault that can extend up to hundreds of miles. However, reverse faults with a dip greater than 45° are usually only a few yards long.

OPPOSITE DIRECTIONS
The northwestward movement of the Pacific Plate and the southeastward movement of the North American Plate cause folds and fissures throughout the region.

Dip angle

3

OBLIQUE-SLIP FAULT
This fault has horizontal as well as vertical movements. Thus, the relative displacement between the edges of the fault can be diagonal. In the oldest faults, erosion usually smoothes the differences in the surrounding terrain, but in more recent faults, cliffs are formed. Transform faults that displace mid-ocean ridges are a specific example of oblique-slip faults.

Elevated block

STRIKE-SLIP FAULT
In this fault the relative movement of the plates is mainly horizontal, along the Earth's surface, parallel to the direction of the fracture but not parallel to the fault plane. Transform faults between plates are usually of this type. Rather than a single fracture, they are generally made up of a system of smaller fractures, slanted from a centerline and more or less parallel to each other. The system can be several miles wide.

Streambeds Diverted by Tectonic Movement

Through friction and surface cracking, a transform fault creates transverse faults and, at the same time, alters them with its movement. Rivers and streams distorted by the San Andreas fault have three characteristic forms: streambeds with tectonic displacement, diverted streambeds, and streambeds with an orientation that is nearly oblique to the fault.

1 Diverted Streambed
The stream changes course as a result of the break.

2 Displaced Streambed
The streambed looks "broken" along its fault line.

WEST COAST OF THE UNITED STATES

Length of California	770 miles (1,240 km)
Length of fault	800 miles (1,300 km)
Maximum width of fault	60 miles (100 km)
Greatest displacement (1906)	20 feet (6 m)

Queen Charlotte Fault

Juan de Fuca Plate

PACIFIC PLATE

San Andreas Fault

NORTH AMERICAN PLATE

San Andreas

East Pacific Ridge

Fault plane

140 years

The average interval between major ruptures that have taken place along the fault. The interval can vary between 20 and 300 years.

PAST AND FUTURE
Some 30 million years ago, the Peninsula of California was west of the present coast of Mexico. Thirty million years from now, it is possible that it may be some distance off the coast of Canada.

Fatal Crack

The great San Andreas fault in the western United States is the backbone of a system of faults. Following the great earthquake that leveled San Francisco in 1906, this system has been studied more than any other on Earth. It is basically a horizontal transform fault that forms the boundary between the Pacific and North American tectonic plates. The system contains many complex lesser faults, and it has a total length of 800 miles (1,300 km). If both plates were able to slide past each other smoothly, no earthquakes would result. However, the borders of the plates are in contact with each other. When the solid rock cannot withstand the growing strain, it breaks and unleashes an earthquake.

Living Water

The water in the oceans, rivers, clouds, and rain is in constant motion. Surface water evaporates, water in the clouds precipitates, and this precipitation runs along and seeps into the Earth. Nonetheless, the total amount of water on the planet does not change. The circulation and conservation of water is driven by the hydrologic, or water, cycle. This cycle begins with evaporation of water from the Earth's surface. The water vapor humidifies as the air rises. The water vapor in the air cools and condenses onto solid particles as microdroplets. The microdroplets combine to form clouds. When the droplets become large enough, they begin to fall back to Earth, and, depending on the temperature of the atmosphere, they return to the ground as rain, snow, or hail.

1 EVAPORATION

Thanks to the effects of the Sun, ocean water is warmed and fills the air with water vapor. Evaporation from humid soil and vegetation increases humidity. The result is the formation of clouds.

TRANSPIRATION

Perspiration is a natural process that regulates body temperature. When the body temperature rises, the sweat glands are stimulated, causing perspiration.

10%

Contribution of living beings, especially plants, to the water in the atmosphere.

THE HUMAN BODY IS 65% WATER.

2 CONDENSATION

In order for water vapor to condense and form clouds, the air must contain condensation nuclei, which allow the molecules of water to form microdroplets. For condensation to occur, the water must be cooled.

FORMATION OF DROPLETS

The molecules of water vapor decrease their mobility and begin to collect on solid particles suspended in the air.

Nucleus

GASEOUS STATE

The rays of the Sun increase the motion of atmospheric gases. The combination of heat and wind transforms liquid water into water vapor.

3 The water vapor escapes via micropores in the leaves' surface.

2 The water ascends via the stem.

1 The root absorbs water.

Root cells

CLOUDS

All the molecules of water are freed.

OCEAN

RIVER

DISCHARGE AREA

6 RETURN TO THE OCEAN

The waters return to the ocean, completing the cycle, which can take days for surface waters and years for underground waters.

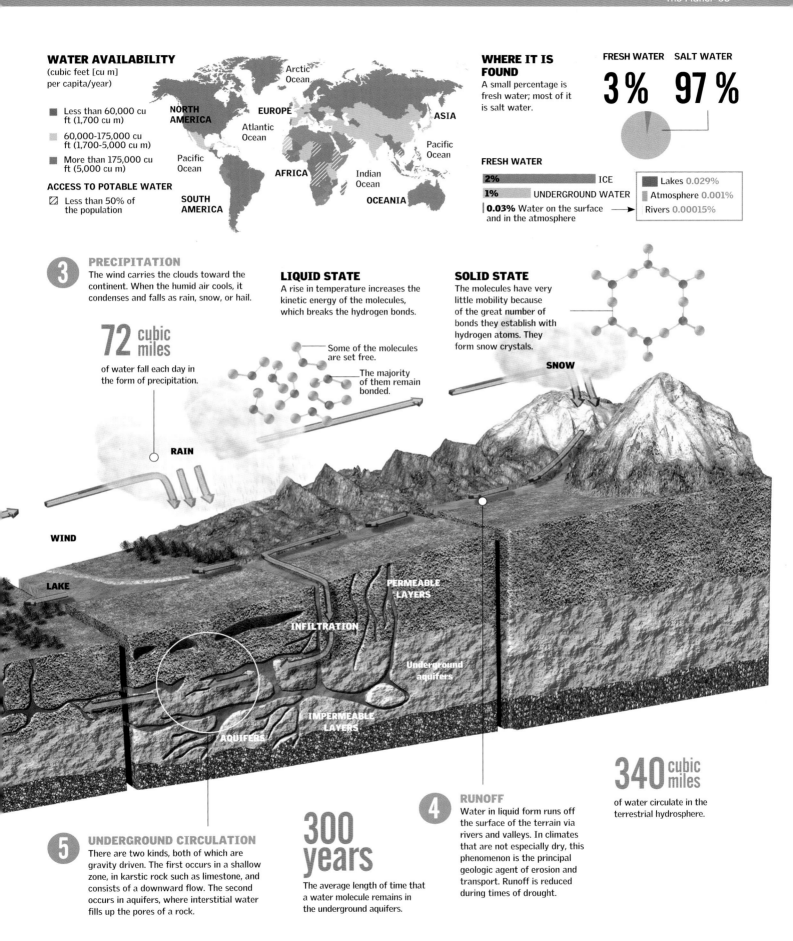

WATER AVAILABILITY
(cubic feet [cu m] per capita/year)

- Less than 60,000 cu ft (1,700 cu m)
- 60,000-175,000 cu ft (1,700-5,000 cu m)
- More than 175,000 cu ft (5,000 cu m)

ACCESS TO POTABLE WATER
- Less than 50% of the population

Arctic Ocean

NORTH AMERICA
Atlantic Ocean
EUROPE
ASIA
Pacific Ocean
Pacific Ocean
AFRICA
Indian Ocean
SOUTH AMERICA
OCEANIA

WHERE IT IS FOUND
A small percentage is fresh water; most of it is salt water.

FRESH WATER **SALT WATER**

3 % 97 %

FRESH WATER

2%	ICE
1%	UNDERGROUND WATER
0.03% Water on the surface and in the atmosphere	

- Lakes 0.029%
- Atmosphere 0.001%
- Rivers 0.00015%

3 PRECIPITATION
The wind carries the clouds toward the continent. When the humid air cools, it condenses and falls as rain, snow, or hail.

LIQUID STATE
A rise in temperature increases the kinetic energy of the molecules, which breaks the hydrogen bonds.

SOLID STATE
The molecules have very little mobility because of the great number of bonds they establish with hydrogen atoms. They form snow crystals.

72 cubic miles
of water fall each day in the form of precipitation.

Some of the molecules are set free.

The majority of them remain bonded.

SNOW

RAIN

WIND

LAKE

PERMEABLE LAYERS

INFILTRATION

Underground aquifers

IMPERMEABLE LAYERS

AQUIFERS

340 cubic miles
of water circulate in the terrestrial hydrosphere.

5 UNDERGROUND CIRCULATION
There are two kinds, both of which are gravity driven. The first occurs in a shallow zone, in karstic rock such as limestone, and consists of a downward flow. The second occurs in aquifers, where interstitial water fills up the pores of a rock.

300 years
The average length of time that a water molecule remains in the underground aquifers.

4 RUNOFF
Water in liquid form runs off the surface of the terrain via rivers and valleys. In climates that are not especially dry, this phenomenon is the principal geologic agent of erosion and transport. Runoff is reduced during times of drought.

Ocean Currents

Ocean water moves as waves, tides, and currents. There are two types of currents: surface and deep. The surface currents, caused by the wind, are great rivers in the ocean. They can be some 50 miles (80 km) wide. They have a profound effect on the world climate because the water warms up near the Equator, and currents transfer this heat to higher latitudes. Deep currents are caused by differences in water density.

The influence of the winds

TIDES AND THE CORIOLIS EFFECT

The Coriolis effect, which influences the direction of the winds, drives the displacement of marine currents.

Currents in the Northern Hemisphere travel in a clockwise direction.

In the Southern Hemisphere, the currents travel in a counterclockwise direction.

GEOSTROPHIC BALANCE

The deflection caused by the Coriolis effect on the currents is compensated for by pressure gradients between cyclonic and anticyclonic systems. This effect is called geostrophic balance.

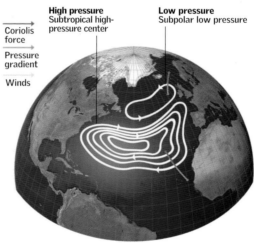

→ Coriolis force
→ Pressure gradient
→ Winds

High pressure Subtropical high-pressure center

Low pressure Subpolar low pressure

North Pacific Current

Pacific Ocean

California Current

Gulf Stream

North Equatorial Countercurrent

Equatorial Countercurrent

Atlantic Ocean

Canary

South Equatorial Current

North Equatorial Countercurrent

Equatorial Cou

Sou

Peruvian Current

Brazil Current

Pacific Ocean

Falkland Current

Antarctic Circumpolar Current

How currents are formed

Wind and solar energy produce surface currents in the water.

1 In the Southern Hemisphere, coastal winds push away the surface water so that cold water can ascend.

2 This slow ascent of deep water is called a surge. This motion is modified by the Ekman spiral effect.

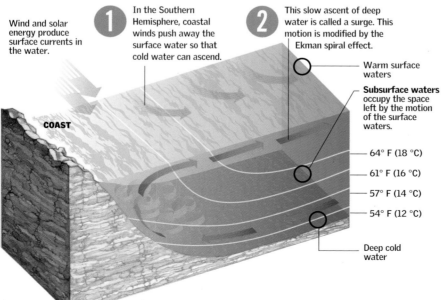

COAST

Warm surface waters

Subsurface waters occupy the space left by the motion of the surface waters.

— 64° F (18 °C)

— 61° F (16 °C)

— 57° F (14 °C)

— 54° F (12 °C)

Deep cold water

EKMAN SPIRAL

explains why the surface currents and deep currents are opposite in direction.

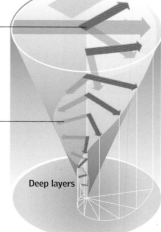

Wind energy is transferred to the water in friction layers. Thus, the velocity of the surface water increases more than that of the deep water.

The Coriolis effect causes the direction of the currents to deviate. The surface currents travel in the opposite direction of the deep currents.

Deep layers

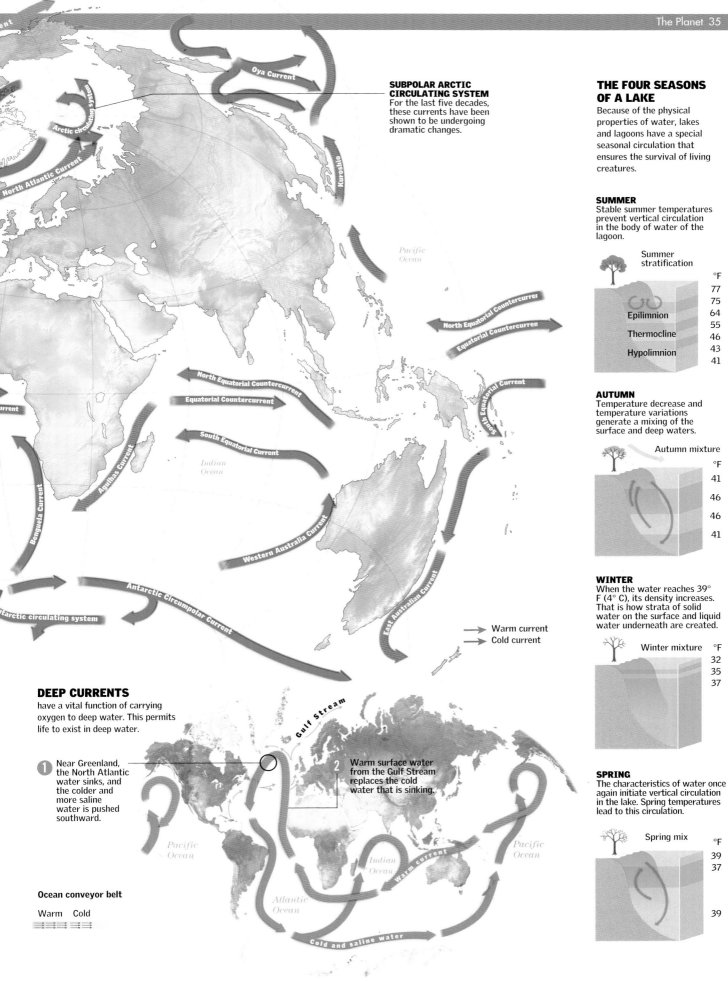

SUBPOLAR ARCTIC CIRCULATING SYSTEM
For the last five decades, these currents have been shown to be undergoing dramatic changes.

Oya Current

Arctic circulating system

North Atlantic Current

Kuroshio

Pacific Ocean

North Equatorial Countercurrent

Equatorial Countercurren

North Equatorial Countercurrent

Equatorial Countercurrent

South Equatorial Current

South Equatorial Current

Indian Ocean

Current

Agulhas Current

Benguela Current

Western Australia Current

East Australian Current

Antarctic Circumpolar Current

Antarctic circulating system

→ Warm current
→ Cold current

DEEP CURRENTS

have a vital function of carrying oxygen to deep water. This permits life to exist in deep water.

Gulf Stream

1 Near Greenland, the North Atlantic water sinks, and the colder and more saline water is pushed southward.

2 Warm surface water from the Gulf Stream replaces the cold water that is sinking.

Pacific Ocean

Indian Ocean

Pacific Ocean

Atlantic Ocean

Warm current

Cold and saline water

Ocean conveyor belt

Warm   Cold

THE FOUR SEASONS OF A LAKE

Because of the physical properties of water, lakes and lagoons have a special seasonal circulation that ensures the survival of living creatures.

SUMMER
Stable summer temperatures prevent vertical circulation in the body of water of the lagoon.

Summer stratification

°F

	77
	75
Epilimnion	64
	55
Thermocline	46
Hypolimnion	43
	41

AUTUMN
Temperature decrease and temperature variations generate a mixing of the surface and deep waters.

Autumn mixture

°F

	41
	46
	46
	41

WINTER
When the water reaches 39° F (4° C), its density increases. That is how strata of solid water on the surface and liquid water underneath are created.

Winter mixture   °F

	32
	35
	37

SPRING
The characteristics of water once again initiate vertical circulation in the lake. Spring temperatures lead to this circulation.

Spring mix   °F

	39
	37
	39

Jets of Water

Geysers are intermittent spurts of hot water that can shoot up dozens of yards into the sky. Geysers form in the few regions of the planet with favorable hydrogeology, where the energy of past volcanic activity has left water trapped in subterranean rocks. Days or weeks may pass between eruptions. Most of these spectacular phenomena are found in Yellowstone National Park (U.S.) and in northern New Zealand.

The eruptive cycle

5 **THE CYCLE REPEATS**
When the water pressure in the chambers is relieved, the spurt of water abates, and the cycle repeats. Water builds up again in cracks of the rock and in permeable layers.

4 **SPURTING SPRAY**
The water spurts out of the cone at irregular intervals. The lapse between spurts depends on the time it takes for the chambers to fill up with water, come to a boil, and produce steam.

CONVECTION FORCES
This is a phenomenon equivalent to boiling water.

A
Water cools and sinks back to the interior, where it is reheated.

B
Bubbles of hot gas rise to the surface and give off their heat.

2 **MOUNTING PRESSURE**
The underground chambers fill with water, steam, and gas at high temperatures, and these are then expelled through secondary conduits to the main vent.

1,450 ft
(442 m)

1,500 ft
(457 m)

TALLEST U.S. BUILDING

RECORD HEIGHT

RECORD HEIGHT
In 1904, New Zealand's Waimangu geyser (now inactive) emitted a record-setting spurt of water. In 1903, four tourists lost their lives when they unknowingly came too close to the geyser.

On average, a geyser can expel up to
7,900 gallons
(30,000 l)
OF WATER PER EVENT

3 **BURSTING FORTH**
The water rises by convection and spurts out the main vent to the chimney or cone. The deepest water becomes steam and explodes outward.

The average height reached of the spurt of water is about
148 feet
(45 m)

CRATER

CHIMNEY

TERRACES
These are shallow, quickly drying pools with stair-step sides.

MAIN VENT

RESERVOIR OR CHAMBER

SECONDARY CONDUIT

1 **HEATED WATER**
Thousands of years after the eruption of a volcano, the area beneath it is still hot. The heat rising from the magma chambers warms water that filters down from the soil. In the subsoil, the water can reach temperatures of up to 518° F (270° C), but pressure from cooler water above keeps it from boiling.

Geyser with multiple chambers

Temperatures up to
194°F
(90° C)

HEAT SOURCE
Magma between 2 and 6 miles (3-10 km) deep, at 930-1,110° F (500-600° C).

PRINCIPAL GEOTHERMAL FIELDS

There are some 1,000 geysers worldwide, and 50 percent are in Yellowstone National Park (U.S.).

Umnak Island (U.S.)
Great Geysir (Iceland)
Kamchatka (Russia)
Steamboat Springs/ Beowawe (U.S.)
El Tatio (Chile)
North Island (New Zealand)
YELLOWSTONE (U.S.)

MINERAL SPRINGS

Their water contains many minerals, known since antiquity for their curative properties. Among other substances, they include sodium, potassium, calcium, magnesium, silicon oxide, chlorine, sulfates, and carbonates. They are very helpful for rheumatic illnesses.

Steam Energy
In Iceland, geothermic steam is used not only in thermal spas but also to power turbines that generate most of the country's electricity.

GRAND PRISMATIC SPRING

This spring, in Yellowstone National Park, is the largest hot spring in the United States and the third largest in the world. It measures 246 by 377 feet (75 by 115 m), and it emits about 530 gallons (2,000 l) of water per minute. It has a unique color: red mixed with yellow and green.

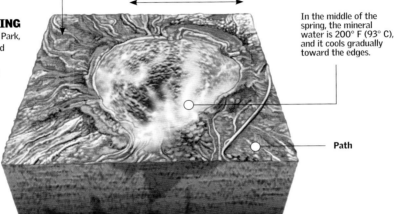

377 feet (115 m)

In the middle of the spring, the mineral water is 200° F (93° C), and it cools gradually toward the edges.

Path

DISCHARGE
530 gallons (2,000 l)
OF WATER PER MINUTE

OTHER POSTVOLCANIC ACTIVITY

FUMAROLE
This is a place where there is a constant emission of water vapor because the temperature of the magma is above 212° F (100° C).

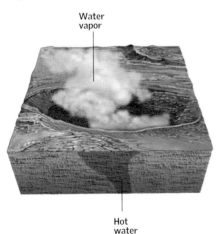

Water vapor

Hot water

SOLFATARA
The thermal layers emit sulfur and sulfurous anhydride.

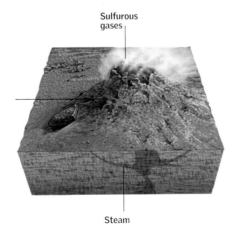

Sulfurous gases

Steam

MUD BASIN
These basins produce their own mud; sulfuric acid corrodes the rocks on the surface and creates a mud-filled hollow.

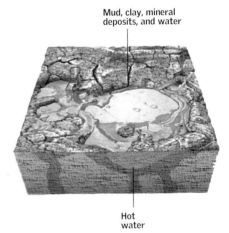

Mud, clay, mineral deposits, and water

Hot water

MORPHOLOGY OF THE CHAMBERS

Great Geysir (Iceland) · Grand Fountain (Yellowstone) · Old Faithful (Yellowstone) · Round Geyser (Yellowstone) · Great Fountain (Yellowstone) · Narcissus (Yellowstone)

Sources of Energy

Energy is vital to life. From it, we get light and heat, and it is what allows economic growth. Most of the energy we use comes from fossil fuels, such as petroleum, coal, and natural gas—substances that took millions of years to form and that will someday be depleted. For this reason, there are more and more countries investing in technologies that take advantage of clean, renewable energy from the Sun, wind, water, and even the interior of the Earth.

NATURAL NUCLEAR REACTOR
The solar energy absorbed by the Earth in a year is equivalent to 20 times the energy stored by all the fossil-fuel reserves in the world and 10,000 times greater than the current consumption of energy.

Nonrenewable Sources

These are the sources of energy that are limited and can forever be depleted through use. They represent up to 85 percent of the world's energy consumption and form the basis of today's insecure energy economy. These nonrenewable sources of energy can be classified into two large groups: fossil fuels (coal, petroleum, and natural gas) and nuclear energy, which is produced in nuclear power plants from uranium—a scarce, controlled radioactive material.

PRIMARY GLOBAL ENERGY SOURCES
Percentages are for the year 2013

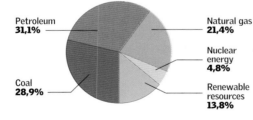

Petroleum **31,1%**
Natural gas **21,4%**
Nuclear energy **4,8%**
Coal **28,9%**
Renewable resources **13,8%**

 A

NUCLEAR ENERGY
B
One of the methods of obtaining electrical energy is through the use of controlled nuclear reaction. This technology continues to be the center of much controversy because of the deadly wastes it generates.

FOSSIL CHEMICAL ENERGY
Fossil fuels (coal, natural gas, and petroleum) are the result of the sedimentation of plants and animals that lived millions of years ago and whose remains were deposited at the bottom of estuaries and swamps. Fossil fuels are the main source of energy for industrial societies. Their combustion releases into the atmosphere most of the gases that cause acid rain and the greenhouse effect.

THERE COULD BE NO MORE COAL RESERVES AFTER THE YEAR

2300

COAL
Coal drove the Industrial Revolution in the developed world. It stiil provides a quarter of the world's commercial energy. Coal is easy to obtain and use, but it is the dirtiest of all energy resources.

GAS MIGHT RUN OUT IN THE YEAR

2150

NATURAL GAS
Formed by the breakdown of organic matter, it can be found in isolation or deposited together with petroleum. One way of transporting it to places of consumption is through gas pipelines.

Renewable Sources

Renewable energy resources are not used up or exhausted through use. As long as they are used wisely, these resources are unlimited because they can be recovered or regenerated. Some of these sources of energy are the Sun, the wind, and water. Depending on the form of exploitation, biomass and geothermal energy can also be considered renewable energy resources.

C HYDROELECTRIC ENERGY
is generated by turbines or water wheels turned by the fall of water. Its main drawback is that the construction of reservoirs, canals, and dams modifies the ecosystems where they are located.

D SOLAR ENERGY
The Sun provides the Earth with great quantities of energy, which can be used for heating as well as for producing electricity.

E WIND ENERGY
ultimately comes from the Sun. Solar radiation creates regions of high and low pressure that creates currents of air in the atmosphere. Wind is one of the most promising renewable energy resources, because it is relatively safe and clean.

F GEOTHERMAL ENERGY
is produced by the heat in the crust and mantle of the Earth. Its energy output is constant, but power plants built to access it must be located in places where water is very close to these heated regions.

G HYDROGEN ENERGY
The production of hydrogen is a new and, for the moment, costly process. But, unlike other fuels, hydrogen does not pollute.

H RENEWABLE CHEMICAL ENERGY BIODIGESTERS
produce fuel from biological resources, such as wood, agricultural waste, and manure. It is the primary source of energy in the developing regions. The methane gas it produces can be used for cooking or to generate electricity.

I TIDAL ENERGY
is one of the newer forms of producing electrical energy. It harnesses the energy released by the ocean as its rises and falls (the ebb and flow of tides).

J BIOFUEL ENERGY
The most common biofuels are ethanol and biodiesel, which are produced from conventional agricultural products, such as oilseeds, sugarcane, or cereals. In the future, they are expected to partially or completely replace gasoline or diesel.

21%
percent of "green" electricity that Europe plans to use in 2010.

PETROLEUM WILL RUN OUT IN THE YEAR

2050

PETROLEUM
Petroleum is the most important energy resource for modern society. If it were to suddenly be depleted, it would be a catastrophe: airplanes, cars, ships, and thermal power plants, among many other things, would be inoperable.

A Changing Surface

The molding of the Earth's crust is the product of two great destructive forces: weathering and erosion. Through the combination of these processes, rocks merge, disintegrate, and join again. Living organisms, especially plant roots and digging animals, cooperate with these geologic processes. Once the structure of the minerals that make up a rock is disrupted, the minerals disintegrate and fall to the mercy of the rain and wind, which erode them.

Erosion

External agents, such as water, wind, air, and living beings, either acting separately or together, wear down, and their loose fragments may be transported. This process is known as erosion. In dry regions, the wind transports grains of sand that strike and polish exposed rocks. On the coast, wave action slowly eats away at the rocks.

HYDROLOGIC PROCESSES

All types of moving water slowly wear down rock surfaces and carry loose particles away. The size of the particles that are carried away from the rock surface depends on the volume and speed of the flowing water. High-volume and high-velocity water can move larger particles.

EOLIAN PROCESSES

The wind drags small particles against the rocks. This wears them down and produces new deposits of either loess or sand depending on the size of the particle.

Wind

River

Weathering

Mechanical agents can disintegrate rocks, and chemical agents can decompose them. Disintegration and decomposition can result from the actions of plant roots, heat, cold, wind, and acid rain. The breaking down of rock is a slow but inexorable process.

CHEMICAL PROCESSES

The mineral components of rocks are altered. They either become new minerals or are released in solution.

Water current

Cave

Limestone

MECHANICAL PROCESSES

A variety of forces can cause rock fragments to break into smaller pieces, either by acting on the rocks directly or by transporting rock fragments that chip away at the rock surface.

TEMPERATURE

When the temperature of the air changes significantly over a few hours, it causes rocks to expand and contract abruptly. The daily repetition of this phenomenon can cause rocks to rupture.

WATER

In a liquid or frozen state, water penetrates into the rock fissures, causing them to expand and shatter.

TRANSPORTATION AND SEDIMENTATION

In this process, materials eroded by the wind or water are carried away and deposited at lower elevations, and these new deposits can later turn into other rocks.

Magmatism

Magma is produced when the temperature in the mantle or crust reaches a level at which minerals with the lowest fusion point begin to melt. Because magma is less dense than the solid material surrounding it, it rises, and in so doing it cools and begins to crystallize. When this process occurs in the interior of the crust, plutonic or intrusive rocks, such as granite, are produced. If this process takes place on the outside, volcanic or effusive rocks, such as basalt, are formed.

Metamorphism

An increase in pressure and/or temperature causes rocks to become plastic and their minerals to become unstable. These rocks then chemically react with the substances surrounding them, creating different chemical combinations and thus causing new rocks to form. These rocks are called metamorphic rocks. Examples of this type of rock are marble, quartzite, and gneiss.

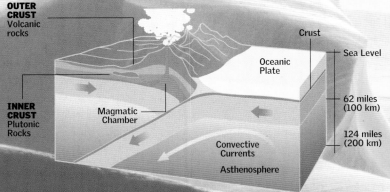

OUTER CRUST
Volcanic rocks

INNER CRUST
Plutonic Rocks

Magmatic Chamber

Crust

Oceanic Plate

Sea Level

62 miles (100 km)

124 miles (200 km)

Convective Currents

Asthenosphere

PRESSURE
This force gives rise to new metamorphic rocks, as older rocks fuse with the minerals that surround them.

TEMPERATURE
High temperatures make the rocks plastic and their minerals unstable.

Folding

Although solid, the materials forming the Earth's crust are elastic. The powerful forces of the Earth place stress upon the materials and create folds in the rock. When this happens, the ground rises and sinks. When this activity occurs on a large scale, it can create mountain ranges or chains. This activity typically occurs in the subduction zones.

FOLDS
For folds to form, rocks must be relatively plastic and be acted upon by a force.

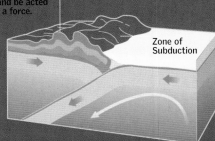

Zone of Subduction

Fracture

When the forces acting upon rocks become too intense, the rocks lose their plasticity and break, creating two types of fractures: joints and faults. When this process happens too abruptly, earthquakes occur. Joints are fissures and cracks, whereas faults are fractures in which blocks are displaced parallel to a fracture plane.

RUPTURE
When rocks rupture quickly, an earthquake occurs.

If Stones Could Speak

Rock strata form from sediments deposited over time in successive layers. Sometimes these sediments bury remains of organisms that can later become fossils, which provide key data about the environment and prehistoric life on Earth. The geologic age of rocks and the processes they have undergone can be discovered through different methods that combine analyses of successive layers and the fossils they contain.

CONTINUITY
The Grand Canyon tells the history of the Earth in colorful layers on its walls. The Colorado River has been carving its way through the plateau for six million years. The layers along the river provide an uninterrupted account of geological history.

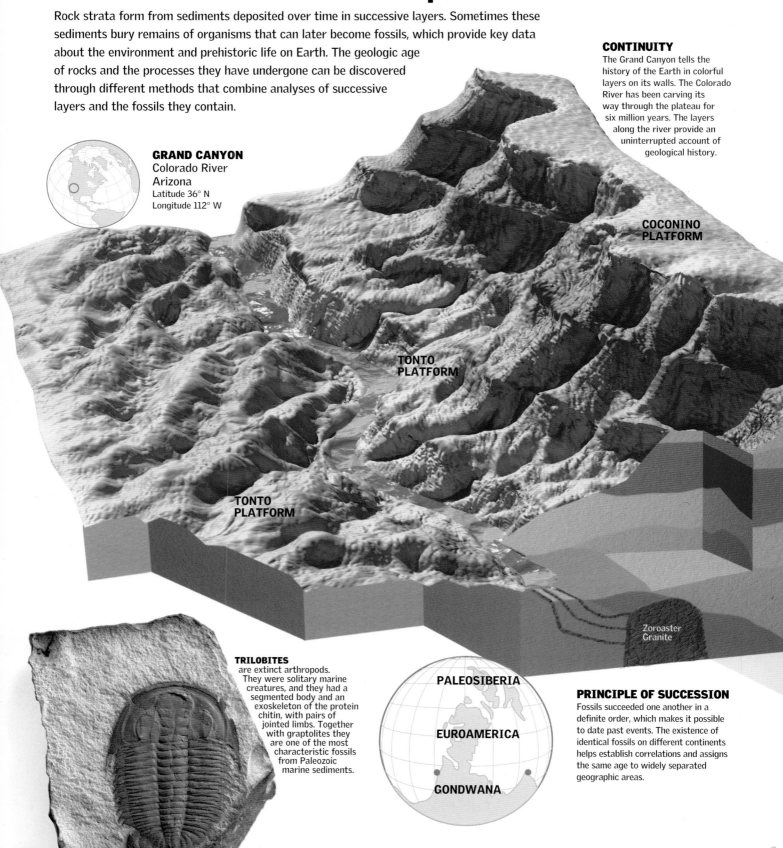

GRAND CANYON
Colorado River
Arizona
Latitude 36° N
Longitude 112° W

COCONINO
PLATFORM

TONTO
PLATFORM

TONTO
PLATFORM

Zoroaster
Granite

TRILOBITES
are extinct arthropods. They were solitary marine creatures, and they had a segmented body and an exoskeleton of the protein chitin, with pairs of jointed limbs. Together with graptolites they are one of the most characteristic fossils from Paleozoic marine sediments.

PALEOSIBERIA

EUROAMERICA

GONDWANA

PRINCIPLE OF SUCCESSION
Fossils succeeded one another in a definite order, which makes it possible to date past events. The existence of identical fossils on different continents helps establish correlations and assigns the same age to widely separated geographic areas.

A Fossil's Age

Fossils are remains of organisms that lived in the past. Today scientists use several procedures, including carbon-14 dating, to estimate their age. This method makes it possible to date organic remains with precision from as long ago as 60,000 years. If organisms are older, there are other methods for absolute dating. However, within a known area, a fossil's location in a given sedimentary layer enables scientists to place it on an efficient, relative time scale. Following principles of original horizontality and of succession, it is possible to find out when an organism lived.

1 When it dies, an animal can be submerged on a riverbed, protected from oxygen. The body begins to decompose.

2 The skeleton is completely covered with sediments. Over the years, new layers are added, burying the earlier layers.

During fossilization, molecules of the original tissue are replaced precisely with minerals that petrify it.

3 Once the water disappears, the fossil is already formed and crystallized. The crust's movements raise the layers, bringing the fossil to the surface.

4 Erosion exposes the fossil to full view. With carbon-14 dating, scientists can determine if it is less than 60,000 years old.

Rock Layers and the Passage of Time

Rock layers are essential for time measurement because they retain information not only about the geologic past but also about past life-forms, climate, and more. The principle of original horizontality establishes that the layers of sediment are deposited horizontally and parallel to the surface and that they are defined by two planes that show lateral continuity. If layers are folded or bent, they must have been altered by some geologic process. These ruptures are called unconformities. If the continuity between layers is interrupted, it means that there was an interval of time and, consequently, erosion in the layer below. This also is called unconformity, since it interrupts the horizontality principle.

Period

PERMIAN

Cononino Sandstone

Hermit Shale

CARBONIFEROUS

Muav Limestone

Bright Angel Shale

DEVONIAN

CAMBRIAN

PRECAMBRIAN

SUPAI GROUP

Paraconformity

REDWALL LIMESTONE

460 feet (140 m)

Disconformity

TONTO GROUP

1,000 feet (310 m)

ANGULAR UNCONFORMITY

Unconformity

UNKAR GROUP

Colorado River

VISHNU SCHISTS

TEMPORAL HIATUS

Unconformity between the Tonto Group and the Redwall Limestone indicates a temporal hiatus. Between the Redwall Limestone and the Supai Group, there is temporal continuity.

Metamorphic Processes

When rocks are subjected to certain conditions (high pressure and temperature or exposure to fluids with dissolved chemicals), they can undergo remarkable changes in both their mineral composition and their structure. This very slow process, called metamorphism, is a veritable transformation of the rock. This phenomenon originates inside the Earth's crust as well as on the surface. The type of metamorphism depends on the nature of the energy that triggers the change. This energy can be heat or pressure.

SLATE
In environments with high temperature and pressure, slates will become phyllites.

Schist

SCOTLAND,
United Kingdom
Latitude   57° N
Longitude 04° W

Dynamic Metamorphism

The least common type of metamorphism, dynamic metamorphism happens when the large-scale movement of the crust along fault systems causes the rocks to be compressed. Great rock masses thrust over others. Where they come in metamorphic rocks, called

2

570° F
(300° C)

SLATE
Metamorphic rock of low grade that forms through pressure at about 390° F (200° C). It becomes more compact and dense.

930° F
(500° C)

SCHIST
Very flaky rock produced by metamorphism at intermediate temperatures and depths greater than six miles (10 km). The minerals recrystallize.

1,200° F
(650° C)

GNEISS
Produced through highly metamorphic processes more than 12 miles (20 km) beneath the surface, it involves extremely powerful tectonic forces and temperatures near the melting point of rock.

1,470° F
(800° C)

FUSION
At this temperature, most rocks start to melt until they become liquid.

Regional Metamorphism

As mountains form, a large amount of rock is deformed and transformed. Rocks buried close to the surface descend to greater depths and are modified by higher temperatures and pressures. This metamorphism covers thousands of square miles and is classified according to the temperature and pressure reached. Slate is an example of rock affected by this type of process.

Contact Metamorphism

Magmatic rocks transmit heat, so a body of magma can heat rocks on contact. The affected area, located around an igneous intrusion or lava flow, is called an aureole. Its size depends on the intrusion and on the magma's temperature. The minerals of the surrounding rock turn into other minerals, and the rock metamorphoses.

1 — Intermediate Crust
— Lower Crust

1 — Sandstone
— Schist
— Limestone
— Magma

2 — Quartzite
— Hornfels
— Marble
— Magma

PRESSURE
As the pressure increases on the rocks, the mineralogical structure of rocks is reorganized, which reduces their size.

TEMPERATURE
The closer the rock is to the heat source and the greater the temperature, the higher the degree of metamorphism that takes place.

Before Rock, Mineral

The planet on which we live can be seen as a large rock or, more precisely, as a large sphere composed of many types of rocks. These rocks are composed of tiny fragments of one or more materials. These materials are minerals, which result from the interaction of different chemical elements, each of which is stable only under specific conditions of pressure and temperature. Both rocks and minerals are studied in the branches of geology called petrology and mineralogy.

Structures

The basic unit of silicates consists of four oxygen ions located at the vertices of a tetrahedron, surrounding a silicon ion. Tetrahedrons can form by sharing oxygen ions, forming simple chains, laminar structures, or complex three-dimensional structures. The structural configuration also determines the type of exfoliation or fracture the silicate will exhibit: mica, which is composed of layers, exfoliates into flat sheets, whereas quartz fractures.

SIMPLE STRUCTURE

All silicates have the same basic component: a silicon-oxygen tetrahedron. This structure consists of four oxygen ions that surround a much smaller silicon ion. Because this tetrahedron does not share oxygen ions with other tetrahedrons, it keeps its simple structure.

COMPLEX STRUCTURE

This structure occurs when the tetrahedrons share three of their four oxygen ions with neighboring tetrahedrons, spreading out to form a wide sheet. Because the strongest bonds are formed between silicon and oxygen, exfoliation runs in the direction of the other bonds, parallel to the sheets. There are several examples of this type of structure, but the most common ones are micas and clays. The latter can retain water within its sheets, which makes its size vary with hydration.

THREE-DIMENSIONAL STRUCTURE

Three fourths of the Earth's crust is composed of silicates with complex structures. Silicas, feldspars, feldspathoids, scapolites, and zeolites all have this type of structure. Their main characteristic is that their tetrahedrons share all their oxygen ions, forming a three-dimensional network with the same unitary composition. Quartz is part of the silica group.

TORRES DEL PAINE
Chilean Patagonia
Latitude 52° 20´ S
Longitude 71° 55´ W

Composition	Granite
Highest summit	Paine Grande (10,000 feet [3,050 m])
Surface	598 acres (242 ha)

Torres del Paine National Park is located in Chile between the massif of the Andes and the Patagonian steppes.

From Minerals to Rocks

From a chemical perspective, a mineral is a homogeneous substance. A rock, on the other hand, is composed of different chemical substances, which, in turn, are components of minerals. The mineral components of rocks are also those of mountains. Thus, according to this perspective, it is possible to distinguish between rocks and minerals.

QUARTZ
Composed of silica, quartz gives rock a white color.

MICA
Composed of thin, shiny sheets of silicon, aluminum, potassium, and other minerals, mica can be black or colorless.

GRANITE
Rock composed of feldspar, quartz, and mica.

FELDSPAR
A light-colored silicate, feldspar makes up a large part of the crust.

12 million years ago
rock batholiths formed during a period of great volcanic activity and created the Torres del Paine and its high mountains.

CHANGE OF STATE
Temperature and pressure play a prominent part in rock transformation. Inside the Earth, liquid magma is produced. When it reaches the surface, it solidifies. A similar process happens to water when it freezes upon reaching 32° F (0° C).

How to Recognize Minerals

A mineral's physical properties are very important for recognizing it at first glance. One physical property is hardness. One mineral is harder than another when the former can scratch the latter. A mineral's degree of hardness is based on a scale, ranging from 1 to 10, that was created by German mineralogist Friedrich Mohs. Another physical property of a mineral is its tenacity, or cohesion—that is, its degree of resistance to rupture, deformation, or crushing. Yet another is magnetism, the ability of a mineral to be attracted by a magnet.

Exfoliation and Fracture

When a mineral tends to break along the planes of weak bonds in its crystalline structure, it separates into flat sheets parallel to its surface. This is called exfoliation. Minerals that do not exfoliate when they break are said to exhibit fracture, which typically occurs in irregular patterns.

TYPES OF EXFOLIATION

Cubic

Octahedral

Dodecahedral

Rhombohedral

Prismatic and Pinacoidal

Pinacoidal (Basal)

TOURMALINE
is a mineral of the silicate group.

COLOR
Some tourmaline crystals can have two or more colors.

FRACTURE
can be irregular, conchoidal, smooth, splintery, or earthy.

IRREGULAR FRACTURE
An uneven, splintery mineral surface

MOHS SCALE

ranks 10 minerals, from the softest to the hardest. Each mineral can be scratched by the one that ranks above it.

1 TALC
is the softest mineral.

2 GYPSUM
can be scratched by a fingernail.

3 CALCITE
is as hard as a bronze coin.

4 FLUORITE
can be scratched by a knife.

5 APATITE
can be scratched by a piece of glass.

Electricity Generation

Piezoelectricity and pyroelectricity are phenomena exhibited by certain crystals, such as quartz, which acquire a polarized charge because exposure to temperature change or mechanical tension creates a difference in electrical potential at their ends.

PIEZOELECTRICITY

The generation of electric currents that can occur when mechanical tension redistributes the negative and positive charges in a crystal. Tourmaline is an example.

PRESSURE

Positive charge

Negative charge

PYROELECTRICITY

The generation of electric currents that can occur when a crystal is subjected to changes in temperature and, consequently, changes in volume.

HEAT

Positive charge

Negative charge

DENSITY

reflects the structure and chemical composition of a mineral. Gold and platinum are among the most dense minerals.

7 to 7.5

is the hardness of the tourmaline on the Mohs scale.

6 **ORTHOCLASE** can be scratched by a drill bit.

7 **QUARTZ** can be scratched by tempered steel.

8 **TOPAZ** can be scratched with a steel file.

9 **CORUNDUM** can be scratched only by diamond.

10 **DIAMOND** is the hardest mineral.

Precious Crystals

Precious stones are characterized by their beauty, color, transparency, and rarity. Examples are diamonds, emeralds, rubies, and sapphires. Compared to other gems, semiprecious stones are composed of minerals of lesser value. Today diamonds are the most prized gem for their "fire," luster, and extreme hardness. The origin of diamonds goes back millions of years, but people began to cut them only in the 14th century. Most diamond deposits are located in South Africa, Namibia, and Australia.

Diamond

Mineral composed of crystallized carbon in a cubic system. The beauty of its glow is due to a very high refraction index and the great dispersion of light in its interior, which creates an array of colors. It is the hardest of all minerals, and it originates underground at great depths.

1 EXTRACTION
Diamonds are obtained from kimberlite pipes left over from old volcanic eruptions, which brought the diamonds up from great depths.

2 CUTTING AND CARVING
The diamond will be cut by another diamond to reach final perfection. This task is carried out by expert cutters.

A INSPECTION:
Exfoliation is determined in order to cut the diamond.

B CUTTING: Using a fine steel blade, the diamond is hit with a sharp blow to split it.

C CARVING:
With a chisel, hammer, and circular saws, the diamond is shaped.

KIMBERLEY MINE

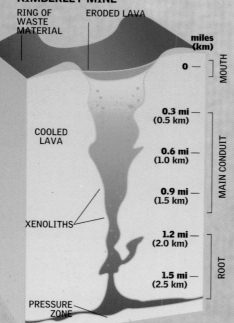

RING OF WASTE MATERIAL
ERODED LAVA

miles (km)

0 — MOUTH

0.3 mi (0.5 km) —
0.6 mi (1.0 km) — MAIN CONDUIT
0.9 mi (1.5 km) —

COOLED LAVA

XENOLITHS

1.2 mi (2.0 km) — ROOT
1.5 mi (2.5 km) —

PRESSURE ZONE

27.6 tons
(25 metric tons)
of mineral must be removed to obtain a 1 carat diamond.

1 carat = 0.007 ounce (0.2 grams)

8 CARATS

0.5 inch (13 mm)

6.5 CARATS

0.3 inch (6.5 mm)

0.03 CARAT

0.08 inch (2 mm)

Gems

Mineral, rock, or petrified material that, after being cut and polished, is used in making jewelry. The cut and number of pieces that can be obtained is determined based on the particular mineral and its crystalline structure.

PRECIOUS STONES

DIAMOND
The presence of any color is due to chemical impurities.

EMERALD
Chromium gives it its characteristic green color.

OPAL
This amorphous silica substance has many colors.

RUBY
Its red color comes from chromium.

3 **POLISHING**
The shaping of the facets of the finished gem.

100
55.1
13.53
34.3°
1.9
40.9°
43.3

CROWN
GIRDLE
PAVILION

IDEAL DIAMOND STRUCTURE

BEZEL
STAR
TABLE

BRILLIANCE
The internal faces of the diamond act as mirrors because they are cut at exact angles and proportions.

FIRE
Flashes of color from a well-cut diamond. Each ray of light is refracted into the colors of the rainbow.

LIGHT — Enters the diamond.

The facets of the pavilion reflect the light among themselves.

The light is reflected back to the crown in the opposite direction.

LIGHT — The rays divide into their components.

Each color reflects separately in the crown.

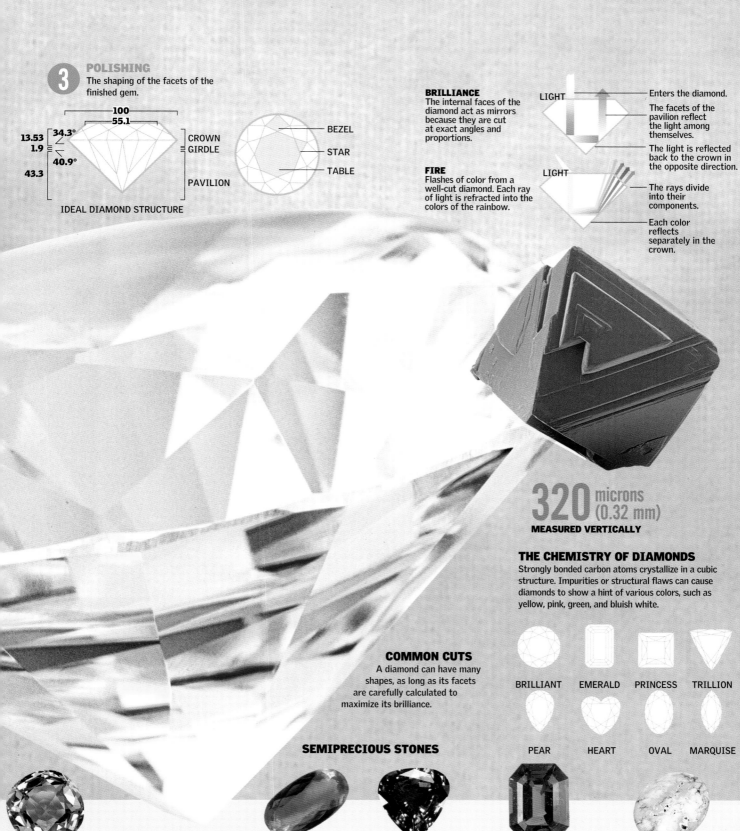

320 microns
(0.32 mm)
MEASURED VERTICALLY

THE CHEMISTRY OF DIAMONDS
Strongly bonded carbon atoms crystallize in a cubic structure. Impurities or structural flaws can cause diamonds to show a hint of various colors, such as yellow, pink, green, and bluish white.

COMMON CUTS
A diamond can have many shapes, as long as its facets are carefully calculated to maximize its brilliance.

BRILLIANT   EMERALD   PRINCESS   TRILLION

PEAR   HEART   OVAL   MARQUISE

SEMIPRECIOUS STONES

SAPPHIRE
Blue to colorless corundum. They can also be yellow.

TOPAZ
A gem of variable color, composed of silicon, aluminum, and fluorine.

AMETHYST
Quartz whose color is determined by manganese and iron.

GARNET
A mix of iron, aluminum, magnesium, and vanadium.

TURQUOISE
Aluminum phosphate and greenish blue copper.

How to Identify Rocks

Rocks can be classified as igneous, metamorphic, or sedimentary according to the manner in which they were formed. Their specific characteristics depend on the minerals that constitute them. Based on this information, it is possible to know how rocks gained their color, texture, and crystalline structure. With a little experience and knowledge, people can learn to recognize and identify some of the rocks that they often see.

Shapes

The final shape that a rock acquires depends to a great extent on its resistance to outside forces. The cooling process and subsequent erosion also influence the formation of rocks. Despite the changes caused by these processes, it is possible to infer information about a rock's history from its shape.

ANGULAR
Rocks have this shape when they have not been worn down.

ROUNDED
The wear caused by erosion and transport gives rocks a smooth shape.

Age

Being able to accurately determine the age of a rock is very useful in the study of geology.

Mineral Composition

Rocks are natural combinations of two or more minerals. The properties of rocks will change in accordance with their mineralogical composition. For instance, granite contains quartz, feldspar, and mica; the absence of any of these elements would result in a different rock.

Color

The color of a rock is determined by the color of the minerals that compose it. Some colors are generated by the purity of the rock, whereas others are produced by the impurities present in it. Marble, for instance, can have different shades if it contains impurities.

WHITE
If the rock is a marble composed of pure calcite or dolomite, it is usually white.

BLACK
Various impurities give rise to different shades in the marble.

0.4 inch (1 cm)

Fracture

When a rock breaks, its surface displays fractures. If the fracture results in a flat surface breaking off, it is called exfoliation. Rocks usually break in locations where their mineral structure changes.

WHITE MARBLE

IMPURITY

WHITE MARBLE

PEGMATITE

WHITE MARBLE

Texture

Refers to the size and arrangement of grains that form a rock. The grains can be thick, fine, or even imperceptible. There are also rocks, such as conglomerates, whose grains are formed by the fragments of other rocks. If the fragments are rounded, there is less compaction, and the rock is therefore more porous. In the case of sedimentary rocks in which the sedimentary cement prevails, the grain is finer.

GRAIN
is the size of the individual parts of a rock, be they crystals and/or fragments of other rocks. A rock's grain can be thick or fine.

0.4 inch (1 cm)

CRYSTALS
form when a melted rock cools and its chemical elements organize themselves. Minerals then take the shape of crystals.

Organic Rocks

Organic rocks are composed of the remains of living organisms that have undergone processes of decomposition and compaction millions of years ago. In these processes, the greater the depth and heat, the greater the caloric power and thermal transformation of the rock. The change experienced by these substances is called carbonization.

Coal Formation

Plant materials, such as leaves, woods, barks, and spores, accumulated in marine or continental basins 285 million years ago. Submerged in water and protected from oxygen in the air, this material slowly became enriched with carbon through the action of anaerobic bacteria.

TRANSFORMATION OF VEGETATION INTO HARD COAL

1 VEGETATION
Organic compounds on the surface became covered by oxygen-poor water found in a peat bog, which effectively shielded them from oxidation.

2 PEAT
Through partial putrefaction and carbonization in the acidic water of the peat bog, the organic matter changes into coal.

Contains **60%** carbon

3 LIGNITE
is formed from the compression of peat that is converted into a brown and flaky substance. Some primary plant structures can still be recognized in it.

Contains **70%** carbon

4 COAL
has a content of less than 40 percent mineral substance on the basis of dry material. It has a matte luster, is similar to charcoal, and is dirty to the touch.

Contains **80%** carbon

5 ANTHRACITE
is the type of coal with the greatest concentration of carbon. Its high heat value is mostly due to this type of coal's high carbon content and low concentration of volatile material. It is harder and denser than ordinary coal.

Contains **95%** carbon

1

2

3

4

5

ANTHRACITE ROCK

At times, the surface of anthracite can appear to have traces of plant fossils.

LOCATION INSIDE THE EARTH

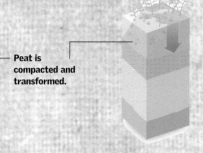

Vegetation that will form peat after dying

The movements of the Earth's crust subjected the strata rich in organic remains to great pressure and transformed them into hard coal over the course of 300 million years.

Peat is compacted and transformed.

DEPTH
up to 1,000 feet
(300 m)

TEMPERATURE
up to 77° F
(25° C)

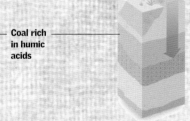

Coal rich in humic acids

DEPTH
1,000 to 5,000 feet
(300 to 1,500 m)

TEMPERATURE
up to 100° F
(40° C)

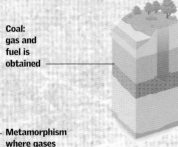

Coal: gas and fuel is obtained

DEPTH
5,000 to 20,000 feet
(1,500 to 6,000 m)

TEMPERATURE
up to 347° F
(175° C)

Metamorphism where gases and oils are released

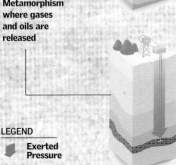

DEPTH
20,000 to 25,000 feet
(6,000 to 7,600 m)

TEMPERATURE
up to 572° F
(300° C)

LEGEND

Exerted Pressure

WORLD COAL RESERVES
Billions of tons

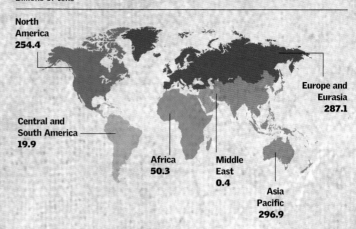

North America
254.4

Europe and Eurasia
287.1

Central and South America
19.9

Africa
50.3

Middle East
0.4

Asia Pacific
296.9

WORLD PETROLEUM RESERVES
Billions of barrels

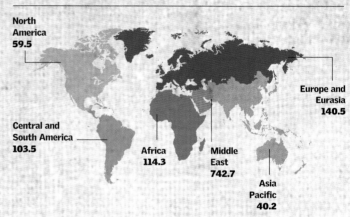

North America
59.5

Europe and Eurasia
140.5

Central and South America
103.5

Africa
114.3

Middle East
742.7

Asia Pacific
40.2

29%
of the primary energy consumed in the world comes from coal.

FORMATION OF PETROLEUM

In an anaerobic environment at a depth of about 1 mile (2 km), organic sediments that developed in environments with little oxygen turn into rocks that produce crude oil.

PETROLEUM TRAPS

Caprock   Storage Rock

ANTICLINE

FAULT TRAP

STRATIGRAPHIC TRAP

SALINE DOME

KEY

Gas

Petroleum (Oil)

Water

ERUPTION
Lava can flow slowly
through cracks—a quiet
eruption—or explode
violently into the air.

CHAPTER 2

VOLCANOES AND EARTHQUAKES

The lithosphere—Earth's rigid outermost shell—is subdivided into tectonic plates undergoing constant change. All this movement is not always evident, until massive amounts of energy build up and devastating natural phenomena like volcanoes and earthquakes occur.

Flaming Furnace

Volcanoes are among the most powerful manifestations of our planet's dynamic interior. The magma they release at the Earth's surface can cause phenomena that devastate surrounding areas: explosions, enormous flows of molten rock, fire and ash that rain from the sky, floods, and mudslides. Since ancient times, human beings have feared volcanoes, even seeing their smoking craters as an entrance to the underworld. Every volcano has a life cycle, during which it can modify the topography and the climate and after which it becomes extinct.

LIFE AND DEATH OF A VOLCANO: THE FORMATION OF A CALDERA

1 Explosive eruptions can expel huge quantities of lava, gas, and rock.

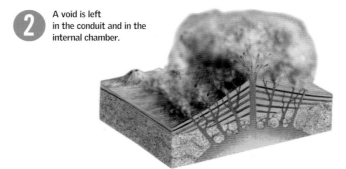

2 A void is left in the conduit and in the internal chamber.

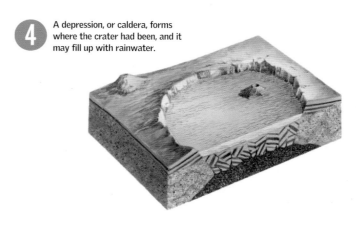

3 The cone breaks up into concentric rings and sinks into the chamber.

Volcanic activity may continue.

4 A depression, or caldera, forms where the crater had been, and it may fill up with rainwater.

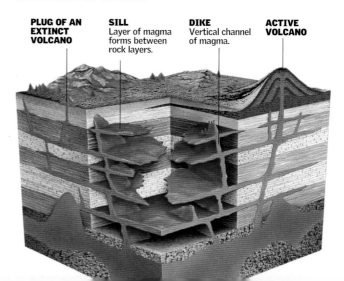

UNDER THE VOLCANO

In its ascent to the surface, the magma may be blocked in various chambers at different levels of the lithosphere.

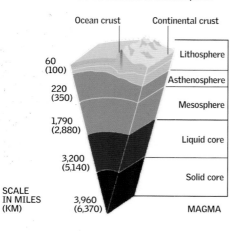

Ocean crust | Continental crust

60 (100)	Lithosphere
220 (350)	Asthenosphere
1,790 (2,880)	Mesosphere
3,200 (5,140)	Liquid core
3,960 (6,370)	Solid core

SCALE IN MILES (KM)

MAGMA

INTRUSION OF MAGMA

PLUG OF AN EXTINCT VOLCANO

SILL Layer of magma forms between rock layers.

DIKE Vertical channel of magma.

ACTIVE VOLCANO

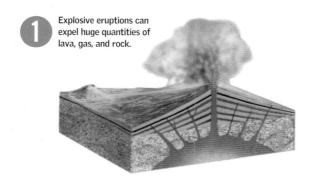

MOUNTAIN-RANGE VOLCANOES

Many volcanoes are caused by phenomena occurring in subduction zones along convergent plate boundaries.

1 When two plates converge, one moves under the other (subduction).

2 The rock melts and forms new magma. Great pressure builds up between the plates.

3 The heat and pressure in the crust force the magma to seep through cracks in the rock and rise to the surface, causing volcanic eruptions.

ERUPTION OF LAVA

CLOUD OF ASH

CRATER
Depression or hollow from which eruptions expel magmatic materials (lava, gas, steam, ash, etc.).

PARASITIC VOLCANO
Composite volcanic cones have more than one crater.

STREAMS OF LAVA
flow down the flanks of the volcano.

VOLCANIC CONE
Made of layers of igneous rock, formed from previous eruptions. Each lava flow adds a new layer.

SECONDARY CONDUIT

MAIN CONDUIT
The pipe through which magma rises. It connects the magma chamber with the surface.

SEEPAGE OF GROUNDWATER

EXTINCT CONDUIT

Magma can reach the surface, or it can stay below ground and exert pressure between the layers of rock. These seepages of magma have various names.

MAGMA CHAMBER
Mass of molten rock at temperatures that may exceed

2,000° F
(1,100° C)

In an active volcano, magma in the chamber is in constant motion because of fluctuations of temperature and pressure (convection currents).

Classification

No two volcanoes on Earth are exactly alike, although they have characteristics that permit them to be studied according to six basic types: shield volcanoes, cinder cones, stratovolcanoes, lava cones, fissure volcanoes, and calderas. A volcano's shape depends on its origin, how the eruption began, and processes that accompany the volcanic activity. They are sometimes classified by the degree of danger they pose to life in surrounding areas.

THE MOST COMMON

Stratovolcanoes, or composite cones, are strung along the edges of the Pacific Plate in the region known as the "Ring of Fire."

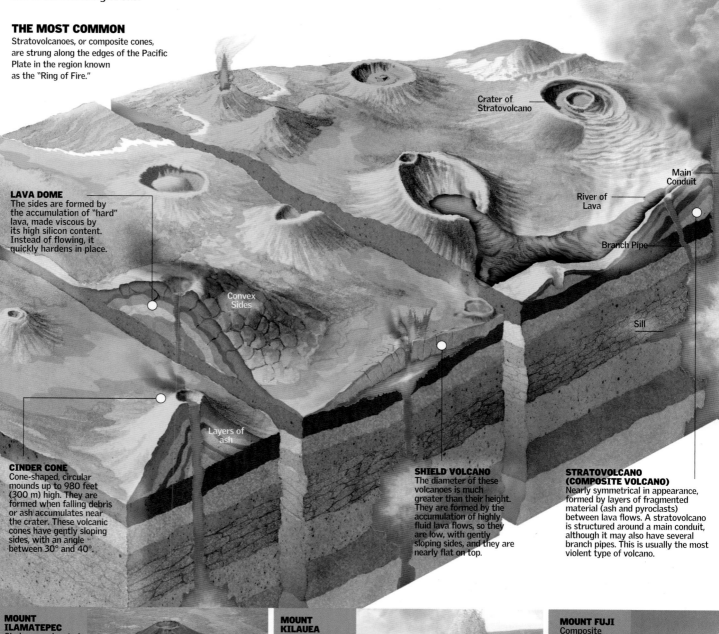

Crater of Stratovolcano

Main Conduit

River of Lava

Branch Pipe

Sill

LAVA DOME

The sides are formed by the accumulation of "hard" lava, made viscous by its high silicon content. Instead of flowing, it quickly hardens in place.

Convex Sides

Layers of ash

CINDER CONE

Cone-shaped, circular mounds up to 980 feet (300 m) high. They are formed when falling debris or ash accumulates near the crater. These volcanic cones have gently sloping sides, with an angle between 30° and 40°.

SHIELD VOLCANO

The diameter of these volcanoes is much greater than their height. They are formed by the accumulation of highly fluid lava flows, so they are low, with gently sloping sides, and they are nearly flat on top.

STRATOVOLCANO (COMPOSITE VOLCANO)

Nearly symmetrical in appearance, formed by layers of fragmented material (ash and pyroclasts) between lava flows. A stratovolcano is structured around a main conduit, although it may also have several branch pipes. This is usually the most violent type of volcano.

MOUNT ILAMATEPEC

Cinder cone located 45 miles (65 km) west of the capital of El Salvador. Its last recorded eruption was in October 2005.

MOUNT KILAUEA

Shield volcano in Hawaii. One of the most active shield volcanoes on Earth.

MOUNT FUJI

Composite volcano 12,400 feet (3,776 m) high, the highest in Japan. Its last eruption was in 1707.

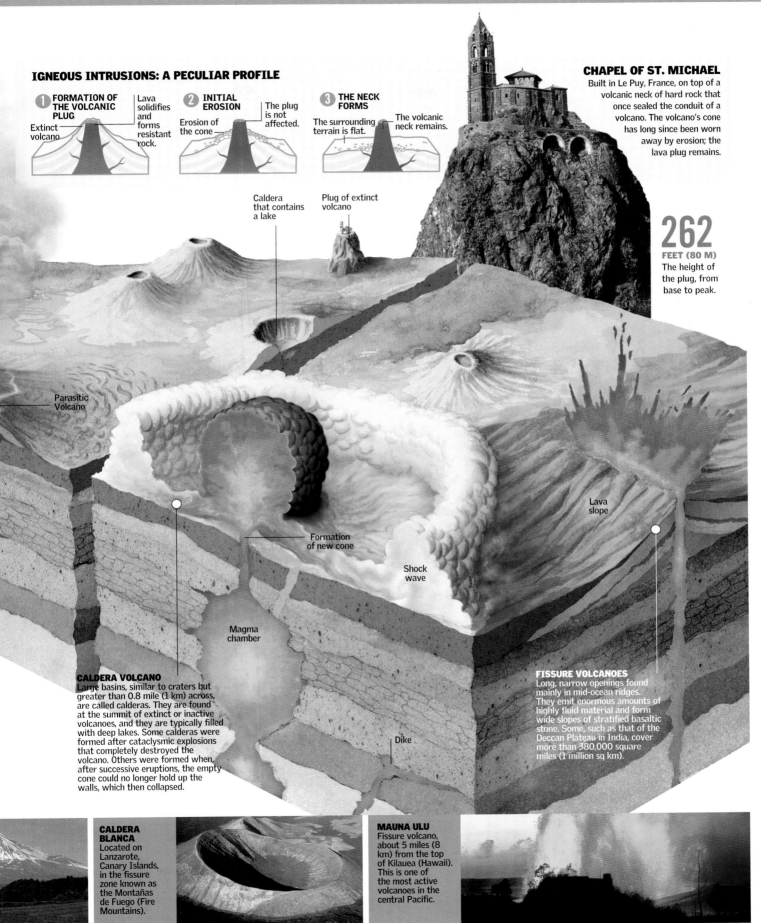

IGNEOUS INTRUSIONS: A PECULIAR PROFILE

1 FORMATION OF THE VOLCANIC PLUG

Extinct volcano

Lava solidifies and forms resistant rock.

2 INITIAL EROSION

Erosion of the cone

The plug is not affected.

3 THE NECK FORMS

The surrounding terrain is flat.

The volcanic neck remains.

CHAPEL OF ST. MICHAEL

Built in Le Puy, France, on top of a volcanic neck of hard rock that once sealed the conduit of a volcano. The volcano's cone has long since been worn away by erosion; the lava plug remains.

262
FEET (80 M)
The height of the plug, from base to peak.

Caldera that contains a lake

Plug of extinct volcano

Parasitic Volcano

Lava slope

Formation of new cone

Shock wave

Magma chamber

Dike

CALDERA VOLCANO
Large basins, similar to craters but greater than 0.8 mile (1 km) across, are called calderas. They are found at the summit of extinct or inactive volcanoes, and they are typically filled with deep lakes. Some calderas were formed after cataclysmic explosions that completely destroyed the volcano. Others were formed when, after successive eruptions, the empty cone could no longer hold up the walls, which then collapsed.

FISSURE VOLCANOES
Long, narrow openings found mainly in mid-ocean ridges. They emit enormous amounts of highly fluid material and form wide slopes of stratified basaltic stone. Some, such as that of the Deccan Plateau in India, cover more than 380,000 square miles (1 million sq km).

CALDERA BLANCA
Located on Lanzarote, Canary Islands, in the fissure zone known as the Montañas de Fuego (Fire Mountains).

MAUNA ULU
Fissure volcano, about 5 miles (8 km) from the top of Kilauea (Hawaii). This is one of the most active volcanoes in the central Pacific.

Flash of Fire

A volcanic eruption is a process that can last from a few hours to several decades. Some are devastating, but others are mild. The severity of the eruption depends on the dynamics between the magma, dissolved gas, and rocks within the volcano. The most potent explosions often result from thousands of years of accumulation of magma and gas, as pressure builds up inside the chamber. Other volcanoes, such as Stromboli and Etna, reach an explosive point every few months and have frequent emissions.

ASH

LAPILLI

HOW IT HAPPENS

3 THE ESCAPE
When the mounting pressure of the magma becomes greater than the materials between the magma and the floor of the volcano's crater can bear, these materials are ejected.

2 IN THE CONDUIT
A solid layer of fragmented materials blocks the magma that contains the volatile gases. As the magma rises and mixes with volatile gases and water vapor, the pockets of gases and steam that form give the magma its explosive power.

1 IN THE CHAMBER
There is a level at which liquefaction takes place and at which rising magma, under pressure, mixes with gases in the ground. The rising currents of magma increase the pressure, hastening the mixing.

4 PYROCLASTIC PRODUCTS
In addition to lava, an eruption can eject solid materials called pyroclasts. Volcanic ash consists of pyroclastic material less than 0.08 inch (2 mm) in size. An explosion can even expel granite blocks.

BOMB	2.5 inches (64 mm) and up
LAPILLI	0.08 to 2.5 inches (2 mm to 64 mm)
ASH	Up to 0.08 inch (2 mm)

CRATER

CONDUIT

Gas Particles   Molten Rock

MAGMA CHAMBER

5 LAVA FLOWS
On the volcanic island of Hawaii, nonerupting flows of lava abound. Local terms for lava include "aa," viscous lava flows that sweep away sediments, and "pahoehoe," more fluid lava that solidifies in soft waves.

EFFUSIVE ACTIVITY

Mild eruptions with a low frequency of explosions. The lava has a low gas content, and it flows out of openings and fissures.

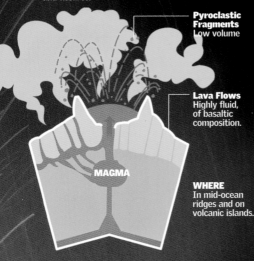

Pyroclastic Fragments Low volume

Lava Flows Highly fluid, of basaltic composition.

MAGMA

WHERE In mid-ocean ridges and on volcanic islands.

EXPLOSIVE ACTIVITY

Comes from the combination of high levels of gas with relatively viscous lava, which can produce pyroclasts and build up great pressure. Different types of explosions are distinguished based on their size and volume. The greatest explosions can raise ash into a column several miles high.

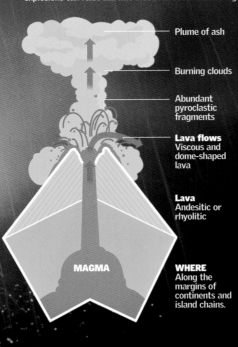

Plume of ash

Burning clouds

Abundant pyroclastic fragments

Lava flows Viscous and dome-shaped lava

Lava Andesitic or rhyolitic

MAGMA

WHERE Along the margins of continents and island chains.

TYPES OF EFFUSIVE ERUPTION

Dome Low, like a shield volcano, with a single opening.

Large, Frequent Lava Flows

Fissure Often several miles long

Lava Seeps out slowly

HAWAIIAN
Volcanoes such as Mauna Loa and Kilauea expel large amounts of basaltic lava with a low gas content, so their eruptions are very mild. They sometimes emit vertical streams of bright lava ("fountains of fire") that can reach up to 330 feet (100 m) in height.

FISSURE
Typical in ocean rift zones, fissures are also found on the sides of composite cones such as Etna (Italy) or near shield volcanoes (Hawaii). The greatest eruption of this type was that of Laki, Iceland, in 1783: 2.9 cubic miles (12 cu km) of lava was expelled from a crack 16 miles (25 km) long.

TYPES OF EXPLOSIVE ERUPTION

Cloud can reach above **82,000 feet (25 km)**

The column can reach a height of **49,000 feet (15 km)**

Cloud of burning material from about **330 to 3,300 feet (100-1,000 m)** high

Lava flow

Burning cloud moving down the slope

Lava plug

STROMBOLIAN
The volcano Stromboli in Sicily, Italy, gave its name to these high-frequency eruptions. The relatively low volume of expelled pyroclasts allows these eruptions to occur approximately every five years.

VULCANIAN
Named after Vulcano in Sicily. As eruptions eject more material and become more explosive, they become less frequent. The 1985 eruption of Nevado del Ruiz expelled tens of thousands of cubic yards of lava and ash.

VESUVIAN
Also called Plinian, the most violent explosions raise columns of smoke and ash that can reach into the stratosphere and last up to two years, as in the case of Krakatoa (1883).

PELEAN
A plug of lava blocks the crater and diverts the column to one side after a large explosion. As with Mt. Pelée in 1902, the pyroclastic flow and lava are violently expelled down the slope in a burning cloud that sweeps away everything in its path.

FROM OUTER SPACE

A photo of the eruption of Mt. Augustine in Alaska, taken by the Landsat 5 satellite hours after the March 27, 1986, eruption.

SMOKE COLUMN

7 Miles
(11.5 Km)
HIGH

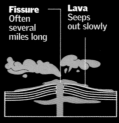

Volcanic ash

Snow and ice

Lava flow

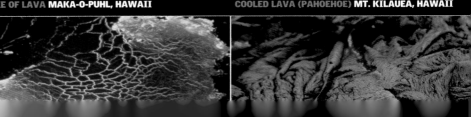

LAVA FLOW **MT. KILAUEA, HAWAII**   LAKE OF LAVA **MAKA-O-PUHL, HAWAII**   COOLED LAVA (PAHOEHOE) **MT. KILAUEA, HAWAII**

Aftermath of Fury

When a volcano becomes active and explodes, it sets in motion a chain of events beyond the mere danger of the burning lava that flows down its slopes. Gas and ash are expelled into the atmosphere and affect the local climate. At times they interfere with the global climate, with more devastating effects. The overflow of lakes can also cause mudslides called lahars, which bury whole cities. In coastal areas, lahars can cause tsunamis.

SNOW

LAVA

VOLCANO

MUD

Lava flows

In volcanoes with calderas, low-viscosity lava can flow without erupting, as with the Laki fissures in 1783. Low-viscosity lava drips with the consistency of clear honey. Viscous lava is thick and sticky, like crystallized honey.

LAVA IN VOLCANO NATIONAL PARK, HAWAII

CINDER CONE

Cone with walls of hardened lava.

As the lava flows upward, the cone explodes.

MOLDS OF TREES

Burned tree underneath cooled lava.

The petrified mold forms a minivolcano.

LAVA TUBES

Outer layer of hardened lava.

Inside, the lava stays hot and fluid.

RESCUE IN ARMERO, COLOMBIA

Mudslide after the eruption of the volcano Nevado del Ruiz. A rescue worker helps a boy trapped in a lahar.

MUDSLIDES OR LAHARS

Rain mixed with snow and melted by the heat, along with tremors and overflowing lakes, can cause mudslides called "lahars." These can be even more destructive than the eruption itself, destroying everything in their path as they flow downhill. They occur frequently on high volcanoes that have glaciers on their summit.

ARMERO FROM ABOVE

On Nov. 13, 1985, the city of Armero, Colombia, was devastated by mudslides from the eruption of the volcano Nevado del Ruiz.

PYROCLASTIC FLOW

Incandescent masses of ash, gas, and rock fragments that come from sudden explosive eruptions flow downhill at high temperature, burning and sweeping away everything in their path.

SPEED
61-132
miles/hour
(100-200 km/h)

TEMPERATURE
930-1830° F
(500-1000° C)

RANGE
30-61
miles/hour
(50-100 km/h)

In rhyolitic eruptions.

Deposit

Nonturbulent dense flow

Turbulent expanded flow

1 Lighter particles separate from heavier ones and rise upward, forming a blanket-shaped cloud.

2 Ahead of the burning cloud, a wave of hot air destroys the forest.

DEADLY FLOW

A bird caught in the eruption of Mount St. Helens, which devastated forests up to a distance of about 8 miles (13 km). The heat and ash left many acres completely destroyed.

AFTEREFFECTS

OPTICAL EFFECTS

Particles of volcanic ash intensify yellow and red colors. After the eruption of Tambora in Indonesia in 1815, unusually colorful sunrises were seen worldwide.

RISING RIVERS

GRAPHICAL RECONSTRUCTION

Aerial illustration of a small fishing village on San Vicente Island, covered in volcanic ash. This eruption had no victims.

QUAKES

The underground action of magma and gas creates pressure that, in turn, causes movement in the Earth's crust. The quakes can be warning signs of an impending eruption.

Latent Danger

Some locations have a greater propensity for volcanic activity. Most of these areas are found where tectonic plates meet, whether they are approaching or moving away from each other. The largest concentration of volcanoes is found in a region of the Pacific known as the "Ring of Fire." Volcanoes are also found in the Mediterranean Sea, in Africa, and in the Atlantic Ocean.

Arctic Ocean

NORTH AMERICA

ASIA

OCEANIA

Pacific Ocean

Indian Ocean

INDO-AUSTRALIAN PLATE

PACIFIC PLATE

The Pacific "Ring of Fire"

Formed by the edges of the Pacific tectonic plate, where most of the world's volcanoes are found.

AVACHINSKY
Russia
This is a young, active cone inside an old caldera, on the Kamchatka peninsula.

NOVARUPTA
Alaska, U.S.
It is in the Valley of Ten Thousand Smokes.

MOUNT ST. HELENS
Washington, U.S.
It had an unexpected, violent eruption in 1980.

FUJIYAMA
Japan
This sacred mountain is the country's largest volcano.

PINATUBO
Philippines
In 1991 it had the second most violent eruption of the 20th century.

50 Volcanoes
Indonesia has the highest concentration of volcanoes in the world. Java alone has 50 active volcanoes.

MAUNA LOA
Hawaii, U.S.
The largest active volcano on Earth is rooted in the ocean floor and takes up nearly half of the island.

KILAUEA
Hawaii, U.S.
The most active shield volcano, its lava flows have covered more than 40 square miles (100 sq km) since 1983.

KRAKATOA
Indonesia
Its 1883 eruption destroyed an entire island.

TAMBORA
Indonesia
In 1815 it produced 35 cubic miles (150 cu km) of ash. It was the largest recorded eruption in human history.

EAST EPI
Vanuatu
This is an undersea caldera with slow eruptions lasting for months.

Subduction
Most volcanoes in the western United States were formed by subduction of the Pacific Plate.

The tallest
These are found in the middle of the Andes range, which forms part of the Pacific Ring of Fire. They were most active 10,000 years ago, and many are now extinct or dampened by fumarolic action.

OJOS DEL SALADO
Chile/Argentina
22,595 ft (6,887 m)

LLULLAILLACO
Chile/Argentina
22,110 ft (6,739 m)

TIPAS
Argentina
21,850 ft (6,660 m)

INCAHUASI
Chile/Argentina
21,720 ft (6,621 m)

SAJAMA
Bolivia
21,460 ft (6,542 m)

MAUNA LOA
Hawaii
Shield volcano
13,680 feet (4,170 m) above sea level.

The "top five" list changes when the volcanoes are measured from the base rather than from their altitude above sea level.

CALDERA

SEA LEVEL

60
volcanoes erupt per year.

Iceland
The western half of Iceland lies on the North American Plate, but the eastern half is on the Eurasian Plate.

EURASIAN PLATE

ELDFELL
Iceland
During one eruption, it expelled 3,500 cubic feet (100 cu m) of lava per second.

ASIA

EUROPE

VESUVIUS
Italy
Erupted twice during the 20th century.

NORTH AMERICAN PLATE

ETNA
Italy
10,990 feet (3,350 m) high; has been active for thousands of years.

The Antilles
The Lesser Antilles is a volcanically active region.

CENTRAL AMERICA

Atlantic Ocean

AFRICA

Indian Ocean

MT. PELÉE
Martinique
Its eruption completely destroyed the city of Saint-Pierre and its port in 1902.

SOUTH AMERICA

NAZCA PLATE

① On May 2, the first rain of ash fell on Saint-Pierre. The sky around the island was darkened for several days.

OJOS DEL SALADO
Chile/Argentina
The tallest volcano in the world, its last eruption was in 1956.

SOUTH AMERICAN PLATE

AFRICAN PLATE

② On May 5, near the summit, the caldera Etang Sec ruptured, releasing the water that it contained. A large lahar formed.

Danger
The most dangerous volcanoes are those located near densely populated areas, such as in Indonesia, the Philippines, Japan, Mexico, and Central America.

③ On May 8, Saint-Pierre was destroyed by a burning cloud that devastated an area of 22 square miles (58 sq km), killing all 28,000 inhabitants.

ANTARCTIC PLATE

Deep Rupture

Earthquakes take place because tectonic plates are in constant motion, and therefore they collide with, slide past, and in some cases even slip on top of each other. The Earth's crust does not give outward signs of all the movement within it. Rather, energy builds up from these movements within its rocks until the tension is more than the rock can bear. At this point the energy is released at the weakest parts of the crust. This causes the ground to move suddenly, unleashing an earthquake.

1 FORESHOCK
Small tremor that can anticipate an earthquake by days or even years. It could be strong enough to move a parked car.

2 AFTERSHOCK
New seismic movement that can take place after an earthquake. At times it can be even more destructive than the earthquake itself.

EARTHQUAKES PER YEAR
30 Seconds
The time lapse between each tremor of the Earth's crust

MAGNITUDE	QUANTITY
8 or Greater	1
7 to 7.9	18
6 to 6.9	120
5 to 5.9	800
4 to 4.9	6,200
3 to 3.9	49,000

EPICENTER
Point on the Earth's surface located directly above the focus.

HYPOCENTER OR FOCUS
Point of rupture, where the disturbance originates. Can be up to 435 miles (700 km) below the surface.

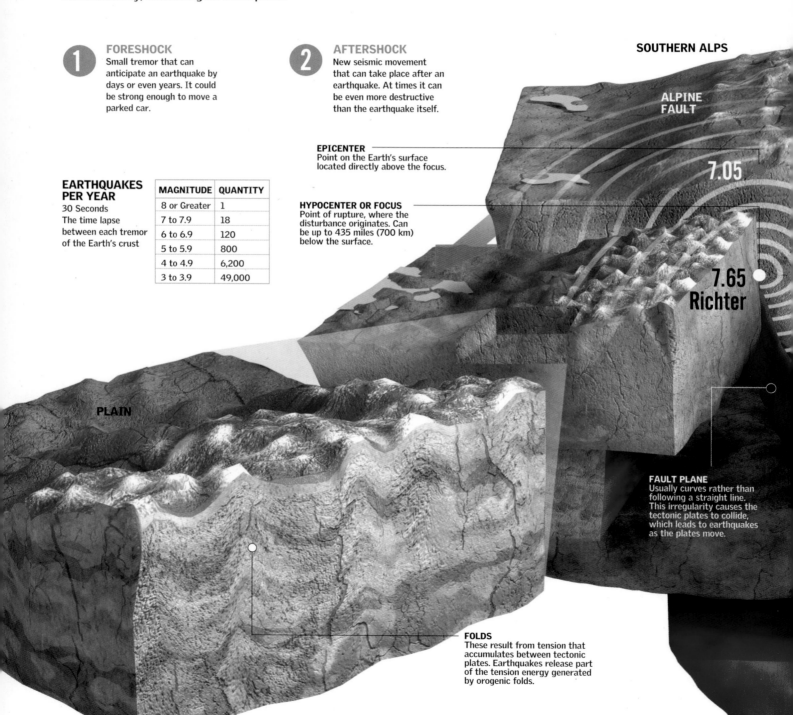

SOUTHERN ALPS

ALPINE FAULT

7.05

7.65 Richter

PLAIN

FAULT PLANE
Usually curves rather than following a straight line. This irregularity causes the tectonic plates to collide, which leads to earthquakes as the plates move.

FOLDS
These result from tension that accumulates between tectonic plates. Earthquakes release part of the tension energy generated by orogenic folds.

ORIGIN OF AN EARTHQUAKE

1 **Tension Is Generated**
The plates move in opposite directions, sliding along the fault line. At a certain point along the fault, they catch on each other. Tension begins to increase between the plates.

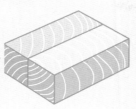

2 **Tension Versus Resistance**
Because the force of displacement is still active even when the plates are not moving, the tension grows. Rock layers near the boundary are distorted and crack.

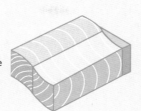

3 **Earthquake**
When the rock's resistance is overcome, it breaks and suddenly shifts, causing an earthquake typical of a transform-fault boundary.

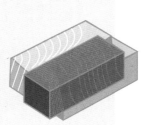

3 **EARTHQUAKE**
The main movement or tremor lasts a few seconds, after which some alterations become visible in the terrain near the epicenter.

NEW ZEALAND
Latitude 42° S
Longitude 174° E

Surface area	103,737 square miles (268,680 sq km)
Population	4,137,000
Population density	35 people per square mile (13.63 people per sq km)
Earthquakes per year (>4.0)	60-100
Total earthquakes per year	14,000

SOUTH ISLAND

LAKE TEKAPO

Riverbeds follow a curved path because of movement along the fault line.

6.10

SEISMIC WAVES transmit the force of the earthquake over great distances in a characteristic back-and-forth movement. Their intensity decreases with distance.

15 miles (25 km)
Average depth of the Earth's crust below the island.

ALPINE FAULT IN NEW ZEALAND

As seen in the cross-section, South Island is divided by a large fault that changes the direction of subduction, depending on the area. To the north the Pacific Plate is sinking under the Indo-Australian Plate at an average rate of 1.7 inches (4.4 cm) per year. To the south, the Indo-Australian Plate is sinking 1.4 inches (3.8 cm) per year under the Pacific Plate.

FUTURE DEFORMATION OF THE ISLAND

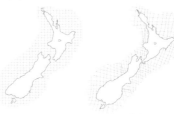

Potential earthquake zone

NORTH ISLAND

Australian Plate

Alpine fault

SOUTH ISLAND

Pacific Plate

To the west there is a plain that has traveled nearly 310 miles (500 km) to the north in the past 20 million years.

2 MILLION YEARS

4 MILLION YEARS

Elastic Waves

Seismic energy is a wave phenomenon, similar to the effect of a stone dropped into a pool of water. Seismic waves radiate out in all directions from the earthquake's hypocenter, or focus. The waves travel faster through hard rock and more slowly through loose sediment and through water. The forces produced by these waves can be broken down into simpler wave types to study their effects.

FOCUS
Vibrations travel outward from the focus, shaking the rock.

2.2 miles per second (3.6 km/s)
S waves are 1.7 times as slow as P waves.
They travel only through solids. They cause splitting motions that do not affect liquids. Their direction of travel is perpendicular to the direction of travel.

Different Types of Waves

There are basically two types of waves: body waves and surface waves. The body waves travel inside the Earth and transmit foreshocks that have little destructive power. They are divided into primary (P) waves and secondary (S) waves. Surface waves travel only along the Earth's surface, but, because of the tremors they produce in all directions, they cause the most destruction.

➡ Direction of seismic waves
➡ Vibration of rock particles

3.7 miles per second (6 km/s)
Typical Speed of P Waves in the Crust.
P waves travel through all types of material, and the waves themselves move in the direction of travel.

Primary Waves

High-speed waves that travel in straight lines, compressing and stretching solids and liquids they pass through.

The ground is **compressed and stretched** by turns along the path of wave propagation.

SPEED IN DIFFERENT MATERIALS

MATERIAL	Granite	Basalt	Limestone	Sandstone	Water
Wave speed in feet per second (m/s)	17,000 (5,200)	21,000 (6,400)	7,900 (2,400)	11,500 (3,500)	4,800 (1,450)

Surface Waves

Appear on the surface after the P and S waves reach the epicenter. Having a lower frequency, surface waves have a greater effect on solids, which makes them more destructive.

1.9 miles per second (3.2 km/s)
Speed of surface waves in the same medium.

These waves travel only along the surface, at 90 percent of the speed of S waves.

RAYLEIGH WAVES
These waves spread with an up-and-down motion, similar to ocean waves, causing fractures perpendicular to their travel by stretching the ground.

LOVE WAVES
These move like horizontal S waves, trapped at the surface, but they are somewhat slower and make cuts parallel to their direction.

The soil is moved to both sides.

The ground is moved in an elliptical pattern.

The soil is moved to both sides, perpendicular to the wave's path of motion.

Secondary Waves

Body waves that shake the rock up and down and side to side as they move.

SPEED IN DIFFERENT MATERIALS

MATERIAL	Granite	Basalt	Limestone	Sandstone
Wave speed in feet per second (m/s)	9,800 (3,000)	1,500 (3,200)	4,430 (1,350)	7,050 (2,150)

Types of Earthquakes

Although earthquakes generally cause all types of waves, some kinds of waves may predominate. This fact leads to a classification that depends on whether vertical or horizontal vibration causes the most movement. The depth of the focus can also affect its destructiveness.

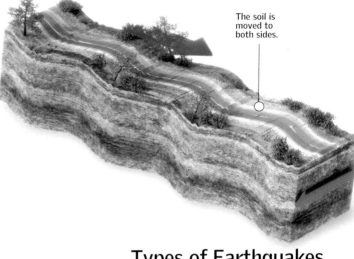

BASED ON TYPE OF MOVEMENT

TREPIDATORY
Located near the epicenter, where the vertical component of the movement is greater than the horizontal.

OSCILLATORY
When a wave reaches soft soil, the horizontal movement is amplified, and the movement is said to be oscillating.

BASED ON FOCUS DEPTH

Earthquakes originate at points between 3 and 430 miles (5 and 700 km) underground. Ninety percent originate in the first 62 miles (100 km). Those originating between 43 and 190 miles (70 and 300 km) are considered intermediate. Superficial earthquakes (often of higher magnitude) occur above that level, and deep-focus earthquakes occur below it.

0
Superficial · 43 miles (70 km)
Intermediate · 190 miles (300 km)
Deep focus · 430 miles (700 km)

TRAJECTORY OF P AND S WAVES

The Earth's outer core acts as a barrier to S waves, blocking them from reaching any point that forms an angle between 105° and 140° from the epicenter. P waves are transmitted farther through the core, but they may be diverted later on.

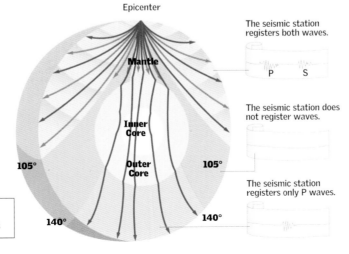

Epicenter

Mantle

Inner Core

Outer Core

105°

105°

140°

140°

The seismic station registers both waves.

P   S

The seismic station does not register waves.

The seismic station registers only P waves.

→ Primary (P) Waves
→ Secondary (S) Waves

Measuring an Earthquake

Earthquakes can be measured in terms of force, duration, and location. Many scientific instruments and comparative scales have been developed to take these measurements. Seismographs measure all three parameters. The Richter scale describes the force or intensity of an earthquake. Naturally, the destruction caused by earthquakes can be measured in many other ways: numbers of people left injured, dead, or homeless, damage and reconstruction costs, government and business expenditures, insurance costs, school days lost, and in many more ways.

CHARLES RICHTER
American seismologist (1900-85) who developed the scale of magnitude that bears his name.

Intensity

Concept of the destruction caused by an earthquake.

Modified Mercalli Scale

Between 1883 and 1902, this Italian volcanologist developed a scale to measure the intensity of earthquakes. It originally had 10 points based on the observation of the effects of seismic activity; it was later modified to 12. The first few levels consist of barely perceptible sensations. The highest levels apply to the destruction of buildings. This scale is widely used to compare levels of damage among different regions and socioeconomic conditions.

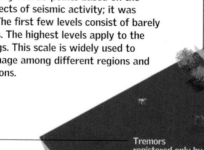

Richter Scale

In 1935, seismologist Charles Richter designed a scale to measure the amplitude of the largest waves registered by seismographs. An important feature of this scale is that the levels increase exponentially. Each point on the scale represents 10 times the movement and 30 times the energy of the point below it. Temblors of magnitude 2 or less are not perceptible to humans. This scale is the most widely used in the world because it can be used to compare the strength of earthquakes apart from their effects.

Magnitude

The energy released in a seismic event.

I	II	III	IV	V
	Hanging objects may swing.	The whole interior of a building vibrates.		Glass windows break.

Trees shake.

The shaking is felt by people inside.

Walls creak.

Windows and doors vibrate.

Tremors registered only by seismographs.

Parked cars rock back and forth.

Animals become upset and anxious.

Church bells sound.

2	2.5	3.5	4.0	5.5
Registered only by seismographs.	Very few people feel the tremor.	The tremor is felt. Only minor damages.	Most people perceive the quake.	Some buildings are lightly damaged.

GIUSEPPE MERCALLI
Italian volcanologist (1850-1914) who developed the first scale for measuring the intensity of an earthquake.

EMS 98 SCALE
In use since 1998 throughout the European Union and other countries that use the protocol, including those of northern Africa. This scale describes the intensity of earthquakes in European contexts, where the most modern construction may be found side by side with ancient buildings. Earthquakes there can have widely varying effects. The scale has 12 points that combine magnitude readings with levels of destruction.

USE OF SCALES WORLDWIDE
○ Richter and Mercalli ● EMS

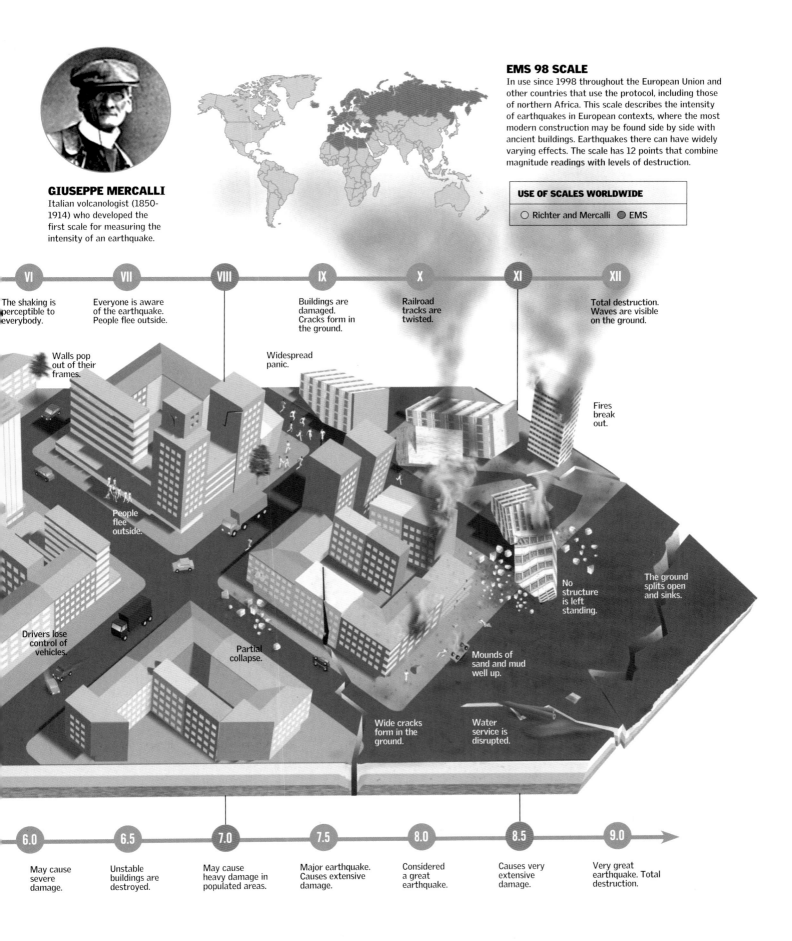

VI The shaking is perceptible to everybody.

VII Everyone is aware of the earthquake. People flee outside.

VIII Buildings are damaged. Cracks form in the ground.

IX Railroad tracks are twisted.

X

XI Total destruction. Waves are visible on the ground.

XII

Walls pop out of their frames.

Widespread panic.

Fires break out.

People flee outside.

Drivers lose control of vehicles.

Partial collapse.

No structure is left standing.

The ground splits open and sinks.

Mounds of sand and mud well up.

Wide cracks form in the ground.

Water service is disrupted.

6.0 May cause severe damage.

6.5 Unstable buildings are destroyed.

7.0 May cause heavy damage in populated areas.

7.5 Major earthquake. Causes extensive damage.

8.0 Considered a great earthquake.

8.5 Causes very extensive damage.

9.0 Very great earthquake. Total destruction.

Violent Seas

A large earthquake or volcanic eruption can cause a tsunami, which means "wave in the harbor" in Japanese. Tsunamis travel very fast, up to 500 miles per hour (800 km/h). On reaching shallow water, they decrease in speed but increase in height. A tsunami can become a wall of water more than 33 feet (10 m) high on approaching the shore. The height depends partly on the shape of the beach and the depth of coastal waters. If the wave reaches dry land, it can inundate vast areas and cause considerable damage. A 1960 earthquake off the coast of Chile caused a tsunami that swept away communities along 500 miles (800 km) of the coast of South America. Twenty-two hours later the waves reached the coast of Japan, where they damaged coastal towns.

THE EARTHQUAKE
A movement of the ocean floor displaces an enormous mass of water upward.

How It Happens

A tremor that generates vibrations on the ocean water's surface can be caused by seismic movement on the seafloor. Most of the time the tremor is caused by the upward or downward movement of a block of oceanic crust that moves a mass of ocean water. A volcanic eruption, meteorite impact, or nuclear explosion can also cause a tsunami.

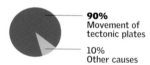

90%
Movement of tectonic plates

10%
Other causes

RISING PLATE

Water level rises   Water level drops

SINKING PLATE

The displaced water tends to level out, generating the force that causes waves.

7.5

Only earthquakes above this magnitude on the Richter scale can produce a tsunami strong enough to cause damage.

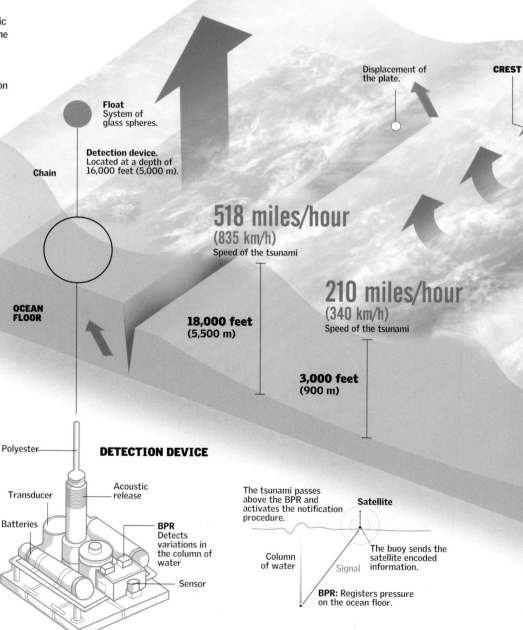

Float
System of glass spheres.

Detection device.
Located at a depth of 16,000 feet (5,000 m).

Chain

OCEAN FLOOR

Displacement of the plate.

CREST

518 miles/hour
(835 km/h)
Speed of the tsunami

210 miles/hour
(340 km/h)
Speed of the tsunami

18,000 feet
(5,500 m)

3,000 feet
(900 m)

Polyester

DETECTION DEVICE

Transducer

Acoustic release

Batteries

BPR
Detects variations in the column of water

Sensor

The tsunami passes above the BPR and activates the notification procedure.

Satellite

The buoy sends the satellite encoded information.

Column of water

Signal

BPR: Registers pressure on the ocean floor.

WHEN THE WAVE HITS THE COAST

A **SEA LEVEL DROPS ABNORMALLY LOW**
Water is "sucked" away from the coast by the growing wave.

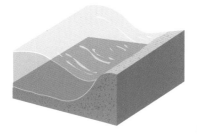

B **THE GIANT WAVE FORMS**
At its highest, the wave may become nearly vertical.

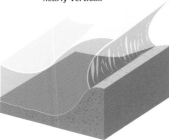

COMPARISON OF THE SIZE OF THE WAVE

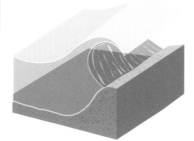

33 feet (10 m)
25 feet (8 m)
9 feet (3 m)
6 feet (1.8 m)

33 feet
(10 m)
Typical height a major tsunami can reach.

C **THE WAVE BREAKS ALONG THE COAST**
The force of the wave is released in the impact against the coast. There may be one wave or several waves.

D **THE LAND IS FLOODED**
The water may take several hours or even days to return to its normal level.

2

THE WAVES ARE FORMED
As this mass of water drops, the water begins to vibrate. The waves, however, are barely 1.5 feet (0.5 m) high, and a boat may cross over them without the crew even noticing.

3

THE WAVES ADVANCE
Waves may travel thousands of miles without weakening. As the sea becomes shallower near the coast, the waves become closer together, but they grow higher.

4

TSUNAMI
On reaching the coast, the waves find their path blocked. The coast, like a ramp, diverts all the force of the waves upward.

TROUGH

CREST

LENGTH OF THE WAVE
From 62 to 430 miles (100 to 700 km) on the open sea, measured from crest to crest.

津
波

The word tsunami comes from Japanese
TSU NAMI
Harbor Wave

Buildings on the coast may be damaged or destroyed.

31 miles/hour
(50 km/h)
Speed of the Tsunami

Between 5 and 30 minutes before the tsunami arrives, the sea level suddenly drops.

65 feet
(20 m)

INDIAN OCEAN

Surface area	**28.3 million square miles (73.4 million sq km)**
Percentage of Earth's surface	**14%**
Percentage of total volume of the oceans	**20%**
Length of plate boundaries (in focus)	**745 miles (1,200 km)**
Countries affected in 2004	**21**

Duration

The tremor lasted between 8 and 10 minutes, one of the longest on record. The waves took six hours to reach Africa, over 5,000 miles (8,000 km) away.

Cause and Effect

On Dec. 26, 2004, an earthquake occurred that measured 9.0 on the Richter scale, the third most powerful earthquake since 1900. The epicenter was 100 miles (160 km) off the west coast of Sumatra, Indonesia. This quake generated a tsunami that pummeled all the coasts of the Indian Ocean. The islands of Sumatra and Sri Lanka suffered the worst effects. India, Thailand, and the Maldives also suffered damage, and there were victims as far away as Kenya, Tanzania, and Somalia, in Africa.

AFRICA

7:58

Local time when the tsunami was unleashed (00:58 universal time).

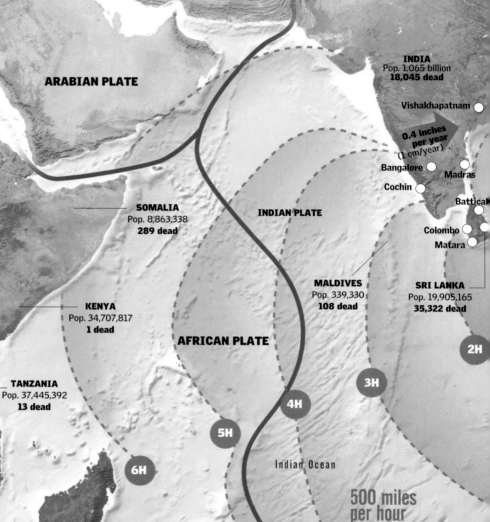

ARABIAN PLATE

INDIA
Pop. 1.065 billion
18,045 dead

Vishakhapatnam

0.4 inches per year
(1 cm/year)

Bangalore

Madras

Cochin

Battical

SOMALIA
Pop. 8,863,338
289 dead

INDIAN PLATE

Colombo

Matara

KENYA
Pop. 34,707,817
1 dead

MALDIVES
Pop. 339,330
108 dead

SRI LANKA
Pop. 19,905,165
35,322 dead

AFRICAN PLATE

2H

TANZANIA
Pop. 37,445,392
13 dead

3H

4H

5H

6H

Indian Ocean

500 miles per hour
(800 km/h)
Speed of the first wave.

THE VICTIMS

On this map, the number of confirmed deaths and the number of missing persons in each country are added together, giving an estimated total death toll. In addition, 1,600,000 persons had to be evacuated.

230,507
Estimated dead

30 per cent
were children

LEGEND

- Most-affected areas
- Plate movements at different speeds
- **6H** Time it took the wave to reach the indicated dotted line.
- Movement of the wave.

BANGLADESH
Pop. 141,340,476
2 dead

EURASIAN PLATE

ASIA

Dhaka

Calcutta

Mandalay

MYANMAR
Pop. 42,720,196
600 dead

Rangoon
(Yangon)

**GULF OF
BENGAL**

Bangkok

THAILAND
Pop. 64,865,523
8,212 estimated dead

PHILIPPINE PLATE

Pacific Ocean

INDONESIA
Pop. 238,452,952
167,736 dead

Phuket

MALAYSIA
Pop. 23,522,482
74 dead

Banda Aceh

4 inches per year
(10 cm/year)

0.4 inches
per year
(1.0 cm/yr)

PACIFIC PLATE

S U M A T R A

0.4 inches
per year
(1.0 cm/yr)

1H

EPICENTER
3° 18' N
95° 47' E

Magnitude 9

Multiple aftershocks of
up to magnitude 7.3.

Sumatra

Banda Aceh

20 sec.

① **Undersea earthquake**
Displacement of 50 feet
(15 m) along the edge
of the Indian Plate, 18
miles (30 km) below the
seabed.

Focus

8 min.

② **The wave begins**
Large waves are
detected northwest and
southeast of the focus.

The tsunami's advance

A seismic station in Australia detected the seismic
movement that later caused the great tsunami that
struck the nearest coastlines with waves more than 33
feet (10 m) high. An hour and a half later, the tsunami
reached Sri Lanka and Thailand. The tsunami had seven
crests, which reached the coasts at 20-minute intervals.
By the time the tsunami arrived at the coast of Africa
hours later, the waves had been greatly diminished.

The wave
reaches land

24 min.

③ **First impact**
A 33-foot-high (10
m) wave destroys
Banda Aceh, Indonesia,
reaching 2.5 miles (4
km) inland.

Risk Areas

A seismic area is found wherever there is an active fault, and these faults are very numerous throughout the world. These fractures are especially common near mountain ranges and mid-ocean ridges. Unfortunately, many population centers were built up in regions near these dangerous places, and, when an earthquake occurs, they become disaster areas. Where the tectonic plates collide, the risk is even greater.

Arctic Ocean

ASIA

6.8
Kobe, 1995
The city of Kobe and nearby villages were destroyed in only 30 seconds.

8.1-8.7
Assam, 1897
More than 1,600 people died in northeast India.

Himalayas

Pacific Ocean

Indo-Australian Plate

MOUNTAIN

TRENCH

Pacific Plate

Subduction zone

PHILIPPINE PLATE

PACIFIC PLATE

MARIANA TRENCH
The deepest marine trench on the planet, with a depth of 35,872 feet (10,934 m) below sea level. It is on the western side of the north Pacific and east of the Mariana Islands.

9.2
Alaska, 1964
Lasted between three and five minutes and caused a tsunami responsible for 122 deaths.

Rocky Mountains

8.3
San Francisco, 1906
Major fires contributed to the devastation of the city.

PACIFIC PLATE

8.1
Mexico, 1985
Two days later there was a 7.6 aftershock. More than 11,000 people died.

Pacific Ocean

9.0
Sumatra, 2004 Tsunami in Asia
An earthquake near the island of Sumatra created 33-foot (10-m) waves and a human tragedy.

Indian Ocean

INDO-AUSTRALIAN PLATE

COCOS AND CARIBBEAN PLATES
Contact between these two plates is of the convergent type: the Cocos Plate moves under the Caribbean Plate, a phenomenon known as subduction. This causes a great number of tremors and volcanoes.

Cocos Plate

Caribbean Plate

Indo-Australian Plate

NEW ZEALAND FAULT
A large fault in which the opposing plates slide past one another; it is a special type of fault called a transform fault.

Pacific Plate

ANTARCTIC PLATE

Most-vulnerable regions

They are unpredictable, and among the most destructive of natural phenomena. Earthquakes shake the earth. They open and move it, and, within a few seconds, they can turn a peaceful city into the worst disaster area, an area in which seismic activity and a high population density coincide. But in the open country, where earthquakes have much less effect, we can conclude that it is not earthquakes, but buildings, that kill people.

EURASIAN PLATE

ASIA

Ural Mountains

EUROPE

Alps

6.8
Armenia, 1988
Destroyed the city of Spitak and took more than 25,000 lives.

7.6
Kashmir, 2005
80,000 fatalities and losses valued at $653,170,000.

8.7
Lisbon, Portugal, 1755
More than 60,000 people died, and a tsunami followed the earthquake.

Caucasus

Himalaya

Atlantic Ocean

NORTH AMERICA

NORTH AMERICAN PLATE

CENTRAL AMERICA

Atlantic Ocean

AFRICA

AFRICAN PLATE

Arabian Plate

7.5
Iran, 1990
More than 60,000 dead. This was the worst disaster in Iran in the 20th century.

ARABIAN PLATE

INDO-AUSTRALIAN PLATE

CARIBBEAN PLATE

COCOS PLATE

African Plate

African Plate

NAZCA PLATE

Andes Mountains

SOUTH AMERICA

SOUTH AMERICAN PLATE

Indian Ocean

AFRICAN AND ARABIAN PLATES

The African Plate includes part of the Atlantic, Indian, and Antarctic oceans. To the north it borders with the Arabian Plate. When these two plates separated, they formed the Red Sea, which is still widening.

MID-OCEAN RIDGES

MID-OCEAN RIDGES

TRENCH

MID-OCEAN RIDGE

A submarine mountain range formed by the displacement of tectonic plates, these are active formations. These mountain systems are the longest in the world.

9.5
Chile, 1960
The most powerful earthquake ever registered: 5,700 people died and two million were left homeless.

South American Plate

Asthenosphere

African Plate

KEY

▲▲▲▲ Convergent boundary

⊢⊢ Oceanic fault

╍╍╍╍ Transform fault

→ Movement and direction of the oceanic fault

⟶ Movement and direction of fault

● Epicentre

🌀 Important earthquake

■ Seismic area

■ Disaster area

ANTARCTIC PLATE

SCOTIA PLATE

ANTARCTIC PLATE

FISH AND CORAL REEF
The depths of the seas and oceans hide fascinating scenes of animal and plant life.

CHAPTER 3

ECOLOGY AND ENVIRONMENT

A branch of biology interested in understanding how organisms relate to one another in their surroundings, ecology's ultimate purpose is to explain that peculiar element that makes the Earth unique: life.

The Six Kingdoms

In order to begin to understand nature, a system must be created to organize the seemingly endless array of organisms. This issue, which has been the topic of many proposals, debates, and disputes among naturalists for centuries, has yet to be completely resolved. Nevertheless, a few methods for classifying organisms have been established that pay attention to the morphological characteristics (or physical features) of different groups and their evolutionary history. Both classification methods are used to determine the relationships between organisms.

Universal Names

There are two types of names for organisms. The common name is the one most people use, but common names can vary from one region to another. The scientific name, derived from Latin, allows any researcher in the world to refer to a specific organism without the possibility of confusing one species with another.

The scientific name of the great white shark is *Carcharodon carcharias.*

Amazon river dolphin, pink dolphin, *boto*, and *bufeo* are common names given to a single animal: *Inia geoffrensis.*

BINOMIAL NOMENCLATURE

By convention, scientific names are generally Latin words; they are written in italics.

Inia geoffrensis

The first word is the genus, and its initial letter is always capitalized.

The second word is a qualifier, and together with the first word it indicates the species.

34

The approximate number of phyla into which the animal kingdom is divided. The mollusk phylum alone (which includes snails, octopuses, and clams) has about 90,000 species.

Classifying Life

At one time, all living things were formally classified into two kingdoms, animal and plant. Today the most widely accepted classification divides organisms into six kingdoms, although a few alternative systems have been proposed and are under discussion.

1 ANIMALIA (ANIMALS)
Multicellular organisms. Their cells are eukaryotic and do not have a cell wall. In general, they are able to move under their own power.

1.5 million

The number of species that have been described by science. It might represent only 5% of all the species in the world.

(2) PLANTAE (PLANTS)

Multicellular organisms. Their cells are eukaryotic, and they have a cell wall. Using a pigment called chlorophyll, they capture energy from sunlight and use it to produce and store their food.

(3) PROTISTA (PROTOZOA)

Unicellular and multicellular eukaryotic organisms that are not part of any other kingdom of life. They include euglenoids, dinoflagellates, fungi, and other eukaryotic microorganisms. (In a eukaryote, the cell's genetic material is organized into chromosomes, and a nuclear membrane separates it from the rest of the cell.)

Paramecium sonneborni Magnified 1,500 times.

(4) EUBACTERIA

Unicellular organisms. They are prokaryotes—that is, they have relatively primitive cells. In prokaryotes, genetic material is not surrounded by a nuclear membrane (as it is in eukaryotes); it is instead inside a cytoplasmic compartment.

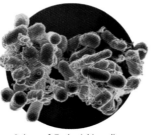

Colony of *Escherichia coli*. Each bacterium is about 100 times smaller than the thickness of a human hair. These bacteria cause various human diseases, such as salmonella.

(5) ARCHAEA

They are a group of unicellular microorganisms. As bacteria, Archaea are prokaryotes that lack a cell nucleus.

(6) FUNGI

Eukaryotic organisms. Traditionally, fungi were included in the plant kingdom, but they now constitute their own group. One of their characteristics is that they form spores. Their cellular structure is very different from that of plants.

Hierarchical Order

Organisms are classified into a system in which some groups are placed within larger groups. For example, domains are divided into kingdoms, which in turn are divided into phyla. Phyla are divided into subphyla and so forth down to the level of species.

An Example: The Classification of Human Beings

Domain:	Eukarya (organisms whose cells contain linear DNA, a cytoskeleton, a nuclear membrane, and other internal membranes).
Kingdom:	Animalia (multicellular organisms that ingest their food).
Phylum:	Chordata (animals that at some time in their life cycle have a hollow dorsal nerve cord and pharyngeal gill slits).
Subphylum:	Vertebrata (animals that have a nerve cord enclosed in a vertebral column).
Superclass:	Tetrapoda (land animals with four limbs).
Class:	Mammalia (the young are nourished with milk from mammary glands; the skin has fur; they are warm-blooded).
Order:	Primates (they have fingers and flat nails, a poor sense of smell, and arboreal habits—or at least their ancestors did).
Family:	Hominidae (bipedal and flat faced, with frontal vision and color vision).
Genus:	Homo (communicates by means of a language). Above, the skulls of three species of the genus Homo are shown.
Species:	Homo sapiens (have a prominent chin, little body hair, and a high forehead).

Homo neanderthalensis

Homo erectus

Homo sapiens

A NEW CLASSIFICATION

The best way of classifying organisms continues to be debated. A new category that has been proposed—the domain—lies above the level of kingdoms. According to this classification scheme, there are three domains (two for prokaryotes and one for eukaryotes), which in turn are divided into kingdoms.

Determining Kinship

Through the study of evolution, it is possible to determine the relatedness and common ancestry of organisms that look very different.

HOMOLOGOUS STRUCTURES

They can be equivalent structures (such as the wings of a bat and the wings of a bird) or different structures (such as the wing of a bird and the arm of a human). Nevertheless, they have a common origin and thus denote a degree of kinship.

ANALOGOUS STRUCTURES

Although these structures appear to be similar or equivalent, a careful analysis will show that they have independent origins (such as the wing of a bird and the wing of an insect). They are simply the result of similar adaptations by organisms to a given environment.

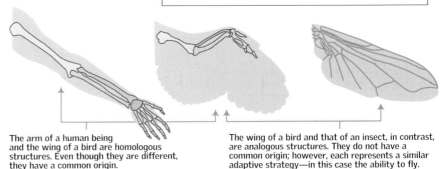

The arm of a human being and the wing of a bird are homologous structures. Even though they are different, they have a common origin.

The wing of a bird and that of an insect, in contrast, are analogous structures. They do not have a common origin; however, each represents a similar adaptive strategy—in this case the ability to fly.

Although they are different, birds and humans are more closely related than birds and insects.

The Basis of Life

Organisms are born, live, reproduce, and die on a natural layer of soil. From this layer, crops are harvested, livestock are raised, and construction materials are obtained. It establishes the link between life and the mineral part of the planet. Through the action of climate and biological agents, soil forms where rocks are broken down.

300 years
The time needed for the natural formation of soil with its three basic layers, or horizons.

Types of Soil

In the soil we find bedrock materials that have been greatly altered by air and water, living organisms, and decomposed organic materials. The many physical and chemical transformations that it undergoes produce different types of soil, some richer in humus, other with more clay, and so on. The soil's basic texture depends to a great extent on the type of bedrock from which the soil is formed.

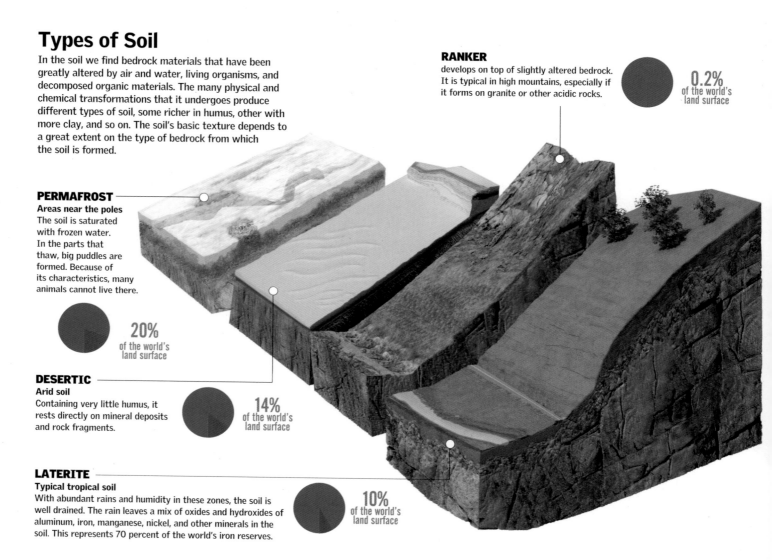

RANKER
develops on top of slightly altered bedrock. It is typical in high mountains, especially if it forms on granite or other acidic rocks.

0.2%
of the world's land surface

PERMAFROST
Areas near the poles
The soil is saturated with frozen water. In the parts that thaw, big puddles are formed. Because of its characteristics, many animals cannot live there.

20%
of the world's land surface

DESERTIC
Arid soil
Containing very little humus, it rests directly on mineral deposits and rock fragments.

14%
of the world's land surface

LATERITE
Typical tropical soil
With abundant rains and humidity in these zones, the soil is well drained. The rain leaves a mix of oxides and hydroxides of aluminum, iron, manganese, nickel, and other minerals in the soil. This represents 70 percent of the world's iron reserves.

10%
of the world's land surface

HOW IT FORMS

Much of the Earth's crust is covered with a layer of sediment and decomposing organic matter. This layer, called soil, covers everything except very steep slopes. Although it is created from decomposing plant and animal remains, the soil is a living and changing system. Its tiniest cavities, home to thousands of bacteria, algae, and fungi, are filled with water or air. These microorganisms speed up the decomposition process, turning the soil into a habitat favorable to plant roots as well as small animals and insects.

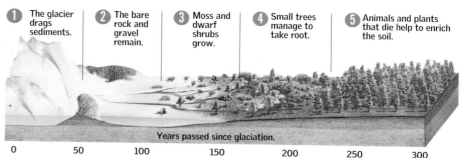

1 The glacier drags sediments.

2 The bare rock and gravel remain.

3 Moss and dwarf shrubs grow.

4 Small trees manage to take root.

5 Animals and plants that die help to enrich the soil.

Years passed since glaciation.

0   50   100   150   200   250   300

Different Characteristics

Observing the soil profile makes it possible to distinguish layers called horizons. Each layer has different characteristics and properties, hence the importance of identifying the layers to study and describe them. The surface layer is rich in organic matter. Beneath is the subsoil, where nutrients accumulate and some roots penetrate. Deeper down is a layer of rocks and pebbles.

0

UPPER LAYER
This layer is dark and rich in nutrients. It contains a network of plant roots along with humus, which is formed from plant and animal residues.

3 ft (1 m)

SUBSOIL
contains many mineral particles from the bedrock. It is formed by complex humus.

7 ft (2 m)

10 ft (3 m)

BEDROCK
The continuous breakdown and erosion of the bedrock helps increase the thickness of the soil. Soil texture also depends to a great extent on the type of bedrock on which it forms.

Living Organisms in the Soil

Many bacteria and fungi live in the soil; their biomass usually surpasses that of all animals living on the surface. Algae (mainly diatoms) also live closest to the surface, where there is most light. Mites, springtails, cochineal insects, insect larvae, earthworms, and others are also found there. Earthworms build tunnels that make the growth of roots easier. Their droppings retain water and contain important nutrients.

EARTHWORMS
It takes approximately 6,000 earthworms to produce 3,000 pounds (1,350 kg) of humus.

HUMUS
is the substance composed of organic materials, usually found in the upper layers of soil. It is produced by microorganisms, mainly acting on fallen branches and animal droppings. The dark color of this highly fertile layer comes from its high carbon content.

Rock Cycle

Some rocks go through the rock cycle to form soil. Under the action of erosive agents, rocks from the Earth's crust take on characteristic shapes. These shapes are a consequence partly of the rock's own composition and partly of several effects caused by erosive agents (meteorological and biological) responsible for breaking down rocky material.

Clouds of dust and ash are released to the atmosphere.

A volcano expels lava and pyroclastic material.

Igneous rock cools down and erodes.

EROSION

Ash and other pyroclastic materials are deposited in layers.

These layers compress and harden.

IGNEOUS ROCK
Extrusive rocks form as the lava cools.

Magma rises to the surface and comes out as lava through the volcano.

Some sedimentary and metamorphic rocks erode, forming new strata.

Igneous and plutonic rocks form as magma cools and solidifies below the Earth's surface.

Heat and pressure can recrystallize the rock without melting it, turning it into another type of rock.

SEDIMENTARY ROCK

METAMORPHIC ROCK

The rock melts to form magma.

If it is hot enough, the rock can turn into magma again.

IGNEOUS ROCK

Ecosystems

Ecosystems include the populations of living things that make up a community and their interactions with the nonliving elements of the environment (the biotope). Although each ecosystem is uniquely variable and complex, all ecosystems exhibit two conditions: (1) a unidirectional flow of energy that originates with the Sun and permits all the various organisms to live and develop and (2) a cyclic flow of various materials. These materials, such as nutrients, originate in the environment, pass through organisms in the environment, and return once again to the environment.

Food Webs and Energy Flows

True food webs are established in each ecosystem. There are primary producers, primary and secondary consumers, and decomposers. The energy flow in such food webs begins with the Sun.

THE SUN
is the principal source of energy on the planet. Life would be impossible without the Sun. Solar energy is used by primary producers (plants and algae) to store chemical energy in the form of sugars.

Every time energy passes from one trophic level to another, there are important losses. Each consumer gains only 10% of the energy that resides in its prey.

PRIMARY PRODUCERS
Plants on land and algae in the water take solar energy and transform it into chemical energy. They make up the first trophic level in the food web.

DECOMPOSERS
These organisms (such as fungi, worms, bacteria, and other microorganisms) are specialized to make use of the sources of energy (such as cellulose and nitrogen compounds) that cannot be used by other animals. Decomposers feed on detritus and other waste products, such as feces and dead animals. As they consume these materials, decomposers return ingredients that circulate in the food web to the environment as new inorganic material.

0.1%

The amount of solar energy reaching the Earth's surface that is used by living things.

PRIMARY CONSUMERS
These consumers are the herbivores that eat the primary producers. Primary consumers require part of the chemical energy that they derive from primary producers to live. Another part is stored within their bodies, and a third part is eliminated without being used.

The Nitrogen Cycle

Nitrogen is a critical element for life. Without it, plants could not live, and thus animals could not exist. Air is made up of 70% nitrogen, but plants cannot use it in its gaseous form. Plants can only take in certain nitrogen compounds found in the soil.

1 Dead animals and the waste products of living animals contain nitrogen that some bacteria and fungi can convert into ammonia (NH_3) and ammonium (NH_4).

2 Other kinds of bacteria convert these compounds into nitrites (NO_2). Nitrites are toxic to plants.

3 Other kinds of bacteria convert some of the nitrites into nitrates (NO_3). Nitrates are absorbed by plants and used for growth.

4 Plant cells also convert the nitrates into ammonium. Ammonium can be combined with carbon to make amino acids, proteins, and other compounds needed by the plants.

5 Animals obtain nitrogen by eating plants. This nitrogen is eventually returned to the soil.

6 Losses: A large portion of the nitrogen is lost from the cycle. Human activity, fire, and water can remove nitrogen from the ecosystem. Some bacteria convert nitrogen in the soil to nitrogen gas, which escapes into the atmosphere.

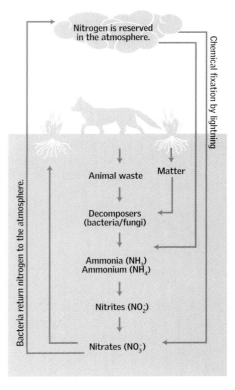

Nitrogen is reserved in the atmosphere.

Chemical fixation by lightning

Bacteria return nitrogen to the atmosphere.

Animal waste   Matter

Decomposers (bacteria/fungi)

Ammonia (NH_3)
Ammonium (NH_4)

Nitrites (NO_2)

Nitrates (NO_3)

The Carbon Cycle

Carbon is a basic constituent of all organic compounds. The most important source of carbon for living organisms is carbon dioxide (CO_2), which makes up almost 0.04% of the air.

1 Carbon dioxide is incorporated into living things by means of plant photosynthesis. Plants use photosynthesis to form organic compounds. In addition, plants expel carbon dioxide through respiration.

2 Herbivores ingest the organic compounds made by plants and reuse them. They also eliminate carbon dioxide through respiration.

3 When a carnivore eats a herbivore, it reuses carbon by incorporating this compound into its body. Carnivores also eliminate carbon dioxide through respiration.

4 The decomposers release carbon dioxide into the atmosphere through respiration.

5 Factories and automobiles burn carbon deposited underground in the form of hydrocarbons and release it as carbon dioxide into the atmosphere.

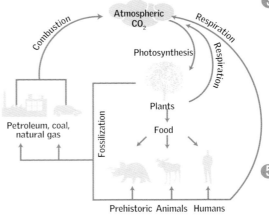

Combustion

Atmospheric CO_2

Respiration

Photosynthesis

Respiration

Petroleum, coal, natural gas

Fossilization

Plants

Food

Prehistoric Animals   Humans

Of all organic material on Earth, 99% corresponds to plants and algae. The rest, in animals, does not exceed 1%.

TERTIARY CONSUMERS
These are the carnivores that eat other carnivores. Some food chains have as many as five trophic levels.

SECONDARY CONSUMERS
These consumers are the carnivores that feed on herbivores. Secondary consumers make use of a small part of the chemical energy stored in primary consumers.

100

species or more can form a food web within an ecosystem.

Not every ecosystem has the Sun as its principal source of energy. Once scientists were able to explore the great depths of the ocean, they began to discover ecosystems whose primary producers are bacteria that use the heat from the Earth's interior as their primary source of energy. These organisms live under extreme conditions, in dark habitats under very high pressures, with temperatures that can exceed 570° F (300° C).

Biodiversity

Contemplating tropical rainforests and coral reefs can be wonderful. At the same time, however, it can be overwhelming because of the abundance and variety of organisms that live in these environments. There are strong indications that the greater the variety of species in an ecosystem, the greater the capacity of the ecosystem to adjust to environmental changes that could potentially threaten it. In addition, research suggests that ecosystems with poor biodiversity are more vulnerable to external factors (such as climate change and invasive species). These observations serve as a wake-up call to human beings, whose activities have disturbed and reduced the biodiversity of the planet.

An Atlas of Variety

It still is not possible to determine how many species inhabit the Earth. What is known is that biodiversity is greatest at or near the tropics. Biodiversity tends to decrease as one approaches the poles.

This map, made at Bonn University, Germany, shows indexes of biodiversity for vascular plants—that is, the majority of the plant kingdom. Some plants (such as algae, mosses, liverworts), however, are not included here.

Species, Genes, and Ecosystems

The concept of biodiversity is typically used to refer to the abundance of species, but it has other meanings in other fields of ecology.

GENETIC DIVERSITY
The genetic characteristics of small, isolated populations in which only a few individuals interbreed are poor when compared to those with larger and more diverse populations. The lack of genetic variability makes a population more susceptible to external changes. Purebred dogs tend to be more "delicate" than most mutts because they are the product of breeding by members of a population with low genetic variability.

THE DIVERSITY OF SPECIES
About 1.5 million species have been described, but the total number of species of living things on the planet is uncertain. As many as 10 million or 100 million species are thought to exist. In addition, it is thought that ecosystems with greater biodiversity are less susceptible to environmental changes than ecosystems with lower biodiversity. In the long run, diverse ecosystems have a higher capacity for recuperation.

THE DIVERSITY OF ECOSYSTEMS
The biosphere is made up of a great number of ecosystems. This diversity gives the biosphere stability and balance and makes it more resilient to major changes. The loss of ecosystems weakens the biosphere and makes it more vulnerable.

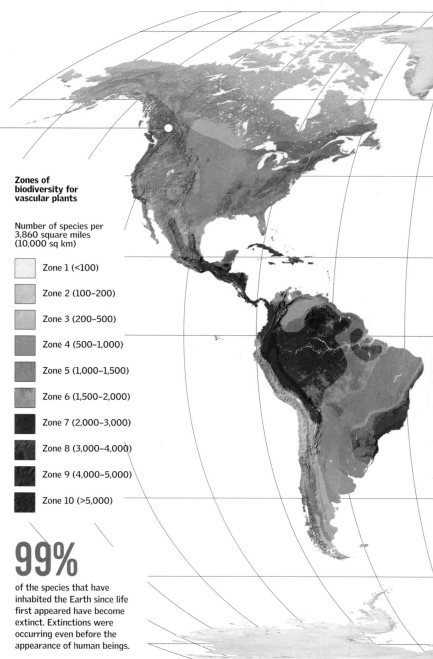

Zones of biodiversity for vascular plants

Number of species per 3,860 square miles (10,000 sq km)

- Zone 1 (<100)
- Zone 2 (100–200)
- Zone 3 (200–500)
- Zone 4 (500–1,000)
- Zone 5 (1,000–1,500)
- Zone 6 (1,500–2,000)
- Zone 7 (2,000–3,000)
- Zone 8 (3,000–4,000)
- Zone 9 (4,000–5,000)
- Zone 10 (>5,000)

99%

of the species that have inhabited the Earth since life first appeared have become extinct. Extinctions were occurring even before the appearance of human beings.

Balanced

Although it is still the subject of heated debate, an increasing number of ecologists maintain that systems with greater diversity are more stable and balanced than those with a meager variety of species.

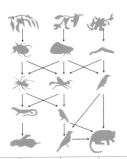

◄ The diagram at left shows an ecosystem of considerable biodiversity with complex food webs.

The ecosystem at right is similar. It is, however, more limited in the sense that its food webs are simpler. The species in this ecosystem have become more vulnerable because they need to depend on fewer resources to survive. In some cases, species depend on a single resource. ►

The Loss of Biodiversity

Human activity is one of the principal factors in the loss of biodiversity today, although it is a process that is difficult to measure. The chart indicates how different factors that are the result of human activity in specific environments affect biodiversity. The present-day trends of these impacts are also shown.

Environment	Change of habitat	Climate change	Over-exploitation	Pollution
Boreal	↗	↑	→	↑
Forest Temperate	↘	↑	→	↑
Tropical	↑	↑	↗	↑
Coastal	↗	↑	↗	↑
Rivers, lakes, and ponds	↑	↑	→	↑

Impact during the 20th century
- Low
- Moderate
- High
- Very high

Current Tendency
- ↘ Decreasing
- → Ongoing
- ↗ Increasing
- ↑ Rapidly increasing

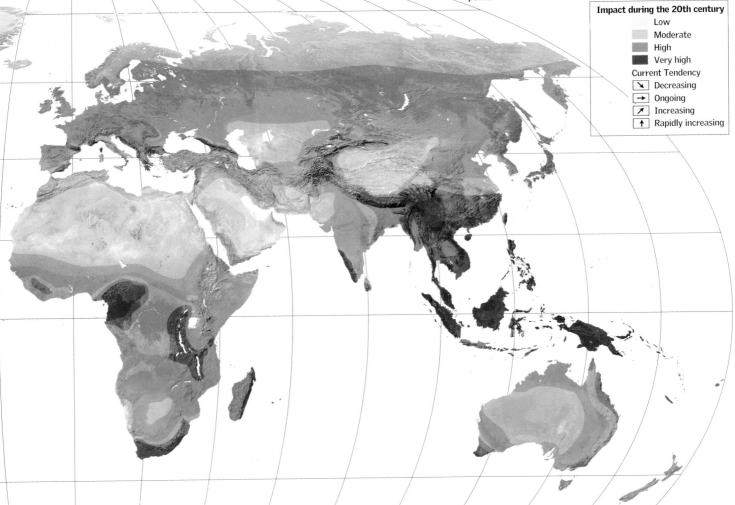

Keystone Species

Some species at the top of the trophic chain are considered "keystone species" because of the effect the loss of this species would have on an ecosystem's biodiversity. One experiment demonstrates that after the elimination of the starfish *Pisaster* in an area inhabited by 15 species, diversity fell to 8 species. This decrease in biodiversity occurred because a dominant species of mussels was allowed to establish itself. These mussels were formerly kept under control by the starfish. *Pisaster* fed on the mussels' prey and therefore left the mussel with limited resources. When the starfish disappeared, the mussel population could freely use these resources. The population of mussels grew, and mussel species outcompeted others.

Pisaster starfish eating a mussel.

Habitats of the World

The great biodiversity that exists on Earth is distributed among specific habitats, the climatic and geologic conditions of which produce specific types of soil, which in turn determine the area's flora and fauna. A biome is a large habitat region that is defined by a characteristic grouping of organisms together with the surrounding environment. Biomes can occur on land or in water.

The Climate Factor

Without a doubt, climate is the most important factor in the distribution of habitats. Such elements as wind, temperature, and precipitation govern the properties of the soil, and therefore the growth of plants, which forms the basis of all biomes. Moisture-laden winds intercepted by mountain ranges produce rainy areas with exuberant forests on one side of the range and arid conditions on the other. Tropical temperatures are a determining factor in the development of coral reefs. Pronounced low winter temperatures can result in the complete loss of tall flora from a habitat, as occurs in the tundra that surrounds the Arctic. The effects of climate, at times beneficial and at other times harmful, drive the complex and varied adaptations of organisms. Some animals have acquired specialized anatomies, whereas others migrate to areas with better conditions.

THE DISTRIBUTION OF HABITATS
Communities of organisms are not distributed arbitrarily. The principal determining factors are air temperature and the availability of water. Temperature decreases with an increase in latitude, and water changes from a liquid to a solid in cold environments. These factors strongly affect the presence of plants and animals and produce very diverse habitats that range from lush tropical rainforests to stark polar and tundra habitats.

NORTHERN HEMISPHERE

SUNLIGHT

SOUTHERN HEMISPHERE

SEASONAL CHANGES
The 23.5° inclination of the Earth's axis, in combination with the Earth's revolution around the Sun, causes the seasonal variation of sunlight in the Northern and Southern Hemispheres.

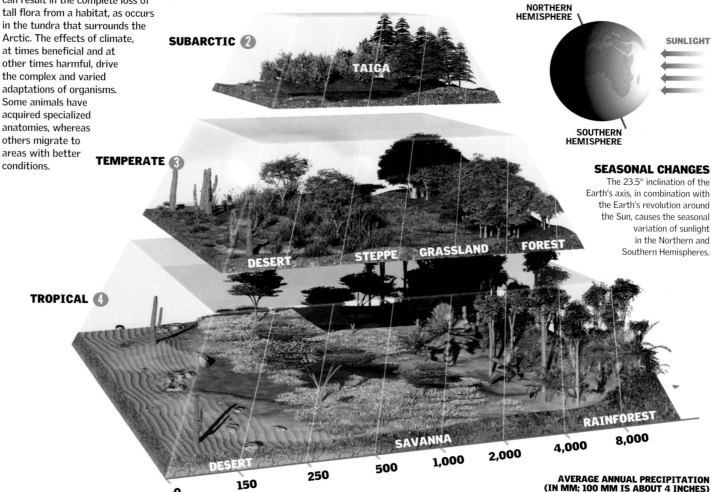

ARCTIC ❶ TUNDRA

SUBARCTIC ❷ TAIGA

TEMPERATE ❸

DESERT STEPPE GRASSLAND FOREST

TROPICAL ❹

RAINFOREST

SAVANNA

DESERT

0   150   250   500   1,000   2,000   4,000   8,000

**AVERAGE ANNUAL PRECIPITATION
(IN MM; 100 MM IS ABOUT 4 INCHES)**

**HABITATS OF
THE WORLD**
The map shows the
distribution and extent
of the principal habitats
of the world, including
terrestrial biomes and marine
environments.

- MOUNTAINS
- DESERTS
- GRASSLANDS
- CONIFEROUS FOREST
- TEMPERATE FOREST
- TROPICAL FOREST
- POLAR REGIONS
- CORAL REEFS

Biodiversity

Every environment has particular characteristics that affect the species that inhabit it. Climate, geology, and other species in the area create a set of conditions that drive the selection of adaptations. These adaptations will determine whether a given species will survive in a given habitat. Skin covered with spines, warm coats of fur, and colorful markings are some of the traits that animals have acquired through the process of natural selection. These traits help species defend themselves against predators, protect themselves from unfavorable climates, or find mates. Among many other specific examples, algae called zoochlorellas live inside the bodies of certain corals and other animals in a mutually beneficial association. In addition, birds such as the oxpecker eat the mites that live on the skin of the water buffalo and other large mammals.

DESERTS
The moloch (*Moloch horridus*), a lizard from Australia, has an armor of sharp, thornlike spines that covers even its head. The protection provided by the spines is essential for this species. The moloch stays in one place for an extended time to feed on ants.

POLAR REGIONS
A thick white coat covers the polar bear (*Ursus maritimus*) and protects it from the extremely cold temperatures of the Arctic. Despite this adaptation, the polar bear is inactive in the winter. It hibernates in a den and survives on its large reserves of body fat.

CORAL REEFS
The spotted clown fish (*Amphiprion* species) lives in the tropical Pacific in association with a sea anemone that allows it to feed and rest among its tentacles. The fish receives protection. It repays this service by keeping the anemone clean. Because both species benefit from this relationship, the interaction is one of mutualism.

TROPICAL FORESTS
As a form of protection, many frogs secrete toxins that can cause paralysis and even death. Brilliant colors are often an indication of the strength of their poison. Such signals allow these frogs to make their way through a forest without being bothered by potential predators. Coloration that serves as a warning is said to be aposematic.

Species Counts

Near the Equator, the Sun passes high overhead all year round, which results in consistently high temperatures. Together with high rainfall, constant high temperatures create the optimum climatic conditions for life. Tropical rainforests are places with the highest concentration of species. The number of species progressively decreases away from the tropics. At the polar ice caps, the lack of biodiversity is compensated for by the large size of animal populations.

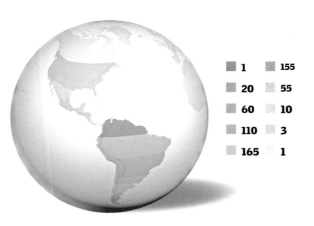

1		155	
20		55	
60		10	
110		3	
165		1	

TROPICAL HUMMINGBIRDS
Hummingbirds are a group of birds native to the Western Hemisphere. The high temperatures and humidity in tropical rainforests foster the vast diversity of these birds. About 150 species of hummingbirds live in Ecuador.

Land Biomes

The different temperatures and moisture levels that affect the earth's surface determine the aptness of soils, which serve as the foundation for sustaining habitats. They also have direct impact on the variety of their animal and plant species, and, therefore, on their ability to adapt in order to survive. The biomes located on solid ground are grouped primarily into forests (temperate, coniferous, and tropical), grasslands, mountains, deserts, and polar regions.

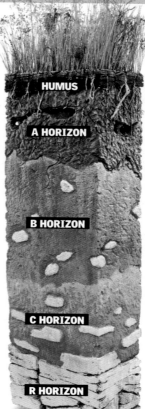

HUMUS

A HORIZON

B HORIZON

C HORIZON

R HORIZON

STRATA
The soil is a key factor in the establishment of each biome. The proportion and composition of the soil's horizons, or layers, directly affects plant growth. The types of plants present in the biome determine the fauna that will live in the biome. Although some soils exhibit every standard horizon, others are made up of only a few. For example, the top layer of humus characteristic of grassland soils is absent in desert regions. In deserts, soil is usually crusted with calcium salts.

Animal Demographics

The distribution of wild animals into distinct populations can occur as the result of natural obstacles or fences and other human-made barriers that cross the land that the animals occupy. A population consists of a group of animals of the same species that interact and occupy a specific place. The characteristic traits of a population are its size, birth rate, death rate, age distribution, and spatial distribution. Two or more species can share the same habitat without competing with each other if their environmental niches do not overlap. This is the case with grazing animals on the savanna.

The Soil

The soil is the foundation of life on Earth. Here, plants extract the mineral nutrients and water from which to grow. Soil is formed by the interactions between rock and the atmosphere, water, and living things. The resulting alteration of the minerals in the bedrock creates an overlying layer of material from which the plants of the various biomes can grow. The various climatic and biological conditions present in a given habitat produce a unique type of soil. Soil itself varies in accordance with the composition of its layers, yielding a distinct mixture of mineral and organic matter.

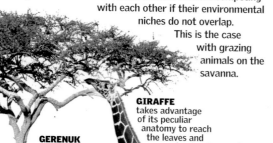

GERENUK
eats the leaves up to 10 feet (3 meters) from the ground.

GIRAFFE
takes advantage of its peculiar anatomy to reach the leaves and new shoots of tall acacias.

ZEBRA
looks for the high, tender shoots of grasses.

GNU
feeds on grass leaves and seed pods.

COOPERATION
The animals grazing on the African savanna do not compete with one another. They complement each other to benefit the plants, which are specially adapted for intense grazing and fires.

DIK-DIKS eat the leaves of shrubs that are no higher than 5 feet (1.5 meters).

GAZELLES AND TOPIS consume the short, dry stalks that are left by the other animals.

Forest Layers

Forest biomes are divided vertically into a number of variable levels, or layers. The principal levels of a temperate forest are the herbaceous, shrub, and tree layers. In a tropical forest, the corresponding layers are the forest floor, the understory, and the canopy. Tropical forests also have an upper level called the emergent layer, which can reach a height of 245 feet (75 meters). There are characteristic flora and fauna for each level. One of the primary factors driving the development of forests is competition for light. This competition manifests itself in the tropical forest canopy by the presence of lianas and epiphytes. At the ground level there is little light, and the vegetative cover is made up of dead leaves and fallen branches. The biomass of this decaying material is equivalent to the biomass that plants of the forest produce in one year. Other organisms, such as fungi and parasitic plants, typically grow under these conditions.

DECIDUOUS FOREST
Trees in this temperate forest lose their leaves in the winter and leaf again in the spring.

CLIMATE
The average annual temperature varies between 75° F and 88° F (24 and 31° C), depending on the altitude, and the relative humidity ranges from 60% to 80%.

SOIL
It is a thick, rich layer of decaying matter that harbors invertebrates and other organisms.

VEGETATION
The hundreds of tree species are veritable living power plants.

Extreme Regions

Climatic factors such as extreme cold and a scarcity of water from precipitation create habitats in which the conditions for life lead to poor vegetation. This lack of high-quality vegetation keeps animal life from becoming established. In the polar regions, the low temperatures reduce the variety of animal species without diminishing their numbers. On the other hand, the water deficit in deserts requires specialized adaptive mechanisms (such as the water-storage capability of cacti). Often, these specialized plants have spines to help increase the efficiency of water conservation.

THE ARCTIC AND TUNDRA
In places where the climate is too cold and the winters too long, the coniferous forest of the Northern Hemisphere gives way to tundra, a vast treeless region that surrounds the Arctic.

CLIMATE
Strong winds reach speeds of 30 to 60 miles (48–96 km) per hour, while the average annual temperature is 7° F (–14° C).

SOIL
Except for places where animals have left droppings, the soil is poor in nutrients and minerals.

VEGETATION
The few species include grasses, mosses, lichens, and sparse shrubs.

DESERT
A desert forms in areas where the evaporation of water exceeds the rainfall.

CLIMATE
Low precipitation—less than 6 inches (about 150 mm) annually. The temperature range is large, as great as 55° F (about 30° C) between night and day.

SOIL
Poorly developed. It has little organic matter in its top layer.

VEGETATION
Notable plants include cacti, which have developed systems for storing water and networks of deep roots to capture water from intermittent rains.

EROSION
Wind, rain, and chemical processes tend to erode the desert plateaus, creating a variety of shapes including deep ravines, isolated hills, arches, and river gullies.

PERMAFROST
The ground in the tundra has underground layers that can remain frozen for more than two years. In the coldest areas, the extent of the permafrost is continuous. It can vary from discontinuous to sporadic in regions where the average temperature is just below 32° F (0° C). The organic matter at the surface gives off greenhouse gases when it thaws.

Continuous   Sporadic
Discontinuous   Isolated

FORESTS
The high density and growth rate of trees—where there is the right amount of light and moisture—give rise to several types of forests in which a complex biological community is adapted to the prevailing temperatures. Depending on the region, a forest may be temperate, boreal, or tropical. Temperate forests are located mainly in the Northern Hemisphere, and tropical forests are located near the Equator. Boreal forests are located below the tundra and are Earth's youngest biome. The trees in temperate forests may be either evergreen or deciduous.

TROPICAL FOREST
Located near the Equator, tropical forests grow constantly and have a great variety of species.

CLIMATE
It is warm and humid all year round.

SOIL
Foliage that falls on the ground forms a layer of putrefying plant matter, which is rapidly mineralized by the conditions. The predominant types of soils, called oxisols, have a characteristic red color because of the presence of iron and aluminum oxides.

VEGETATION
It has the greatest variety of trees, which typically have slender trunks. The forest canopy can reach a height of 245 feet (75 meters). The leaves in the canopy shelter the fauna in the forest below. Only a few palms can grow in the shade of the forest floor.

CONIFEROUS OR BOREAL FOREST
Conifers—trees that reproduce by means of cones—can withstand winter snows, and they form thick, protective forests.

CLIMATE
In the winter, the temperature commonly is –13° F (–25° C), but sometimes reaches –49° F (–45° C).

SOIL
It is acidic as the result of a thick covering of dead tree needles.

VEGETATION
It is limited because of soil acidity and the lack of light that penetrates to the ground.

Aquatic Ecosystems

Water is vital for life. These ecosystems, marine and freshwater, cover a large percentage of the Earth's surface. The oceans contain more than 325 million cubic miles (1,350 million cu km) of water and enough salt to bury Europe to a depth of 3 miles (5 km). The oceans function as enormous boilers that distribute heat from the Sun over the entire planet. The freshwater ecosystems, for their part, supply water for drinking, household use, and irrigation.

Fresh water

Freshwater ecosystems have low concentrations of salt—generally less than 1%. They support 700 species of fish, 1,200 species of amphibians, and a variety of mollusks and insects. Their flora and fauna vary in relation to such factors as the chemical composition of the water, the presence of oxygen, strong water currents, and the timing of periodic droughts. Freshwater ecosystems can be divided into two principal types: standing-water ecosystems (such as lakes, ponds, and wetlands) and flowing-water ecosystems (such as streams and rivers).

SUBMERGED SURFACE

The average depth of the oceans is 12,450 feet (3,795 meters), and their deepest point is the Challenger Deep, in the Mariana Trench, 35,814 feet (10,916 meters) below sea level. That depth exceeds the height of the tallest mountain on the Earth's surface, Mount Everest, which has a height of 29,035 feet (8,850 meters) above sea level.

Greatest depth:
Mariana Trench
(Challenger Deep)
35,814 feet
(10,916 meters)

Average
elevation of
earth's entire
surface:
7,900 feet
(2,400 meters)

Highest point:
Mount Everest
29,035 feet (8,850 meters)

Sea coasts

The meeting of land and sea produces a unique biodiversity, supported by the variety of food present, which is much greater than that of the open ocean. The physical features of the coast are constantly changing because of erosion from ocean waves, coastal currents that carry sand and gravel, and the ebb and flow of the tides. When the tide is low, organisms make adjustments to avoid dehydration. Anemones, for example, draw in their tentacles to become nothing but gelatinous protuberances. Some coastal areas have coral reefs, which are notable for their extraordinary forms and colors. Although beaches of sand, mud, or gravel have fewer ecological resources than do rocky coasts, the former are often used by animals as sites of nest building and shelter from predators.

WATER CYCLE

Water enters an area as precipitation and returns to the atmosphere through evaporation. Some of the water infiltrates the ground. What is left over flows toward bodies of water.

2 PRECIPITATION
As the air continues to condense, it forms droplets of water and flakes of snow.

1 EVAPORATION AND CONDENSATION
Rising air cools, and the water vapor condenses to form clouds.

3 CIRCULATION AND RETURN TO THE SEA
The falling water forms lakes and rivers that discharge into the ocean.

CONSTANT MOTION

Some 270 million years ago, Earth's land surface was joined together into a single continent, Pangea, which was surrounded by water. It later broke into several tectonic plates. As these plates slowly shifted, they progressively reshaped the continents and oceans. Many of the borders between the plates, which continue in constant motion, extend along the ocean bottom. On the borders between two plates, the plates may move away from each other (as in mid-ocean rift zones, where new oceanic crust is formed from rising magma), may move toward each other (as in oceanic trenches, where one plate sinks beneath another), or may slide alongside each other.

The Nature of Sea Salt

The salts in the ocean are composed of a mixture of ions (that is, charged particles) of sodium, chlorine, sulfur, magnesium, potassium, and calcium. Some ions have a positive charge, and others have a negative charge. Some of the salt ions come from rivers, which carry dissolved salts from rocks to the ocean, whereas others come from hydrothermal vents in the ocean floor. A third source of salt ions is from wind-carried volcanic ash. There are different natural processes by which a portion of each type of ion is removed from the ocean, but the most common ions remain dissolved in the ocean. These ions may remain in the ocean from a few centuries up to millions of years.

EASY FLOTATION
In closed seas, such as the Dead Sea, the water is so salty and dense that it makes submersion in the water difficult.

SALT WATER

The average salinity of the ocean is about 35 parts per thousand, although this value varies considerably at the ocean surface. Sources of water, such as rainfall and rivers, reduce the salinity of the ocean, while processes that remove water, such as evaporation, increase it. The freezing point of ocean water depends on its salinity. As salinity increases, the freezing point of water decreases. The salty taste of ocean water is the result of the large amount of chloride it contains.

TYPES OF SALT
The water of the marine ecosystems contains other salts in addition to sodium chloride, such as sulfates and chlorides of magnesium, calcium, and potassium in fixed proportions.

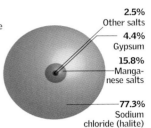

2.5%
Other salts

4.4%
Gypsum

15.8%
Manganese salts

77.3%
Sodium chloride (halite)

UNDERWATER PRESSURE
The pressure of the water increases about one bar for every 33 feet (10 meters) of depth. This fact creates a challenge for undersea scientific explorers, who must undergo slow decompression when they ascend.

Sea level
1 Atmosphere

33 feet (10 meters)
2 Atmospheres

66 feet (20 meters)
3 Atmospheres

Open Ocean

Some organisms live on or just under the surface of the ocean, which has optimal conditions of light and temperature. These organisms are called the neuston. Below the surface, the ocean is divided into four zones—the photic, aphotic, abyssal, and hadal. Each zone corresponds to levels of light and temperature, and these determine the specific kinds of adaptations we see in animals. The most difficult conditions are in the aphotic and abyssal zones because food becomes scarce. In addition, animals in these zones need to retain liquid within their bodies to keep them from being compressed. At depths below about 20,000 feet (6,000 meters), there are ocean trenches that contain hydrothermal vents and strange fish that can withstand the high pressure of the surrounding water.

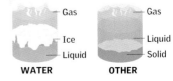

- Gas
- Ice
- Liquid

WATER

- Gas
- Liquid
- Solid

OTHER

DENSITY
Contrary to what might be expected, liquid water is more dense than ice (solid water). Water vapor (a gas) is less dense.

Coral Reefs

Present for more than 450 million years, coral reefs sit beneath shallow, warm ocean waters between the Tropics of Cancer and Capricorn. Coral reefs are veritable rainforests of the sea; their biodiversity encompasses almost one-third of all ocean species. The reefs are solid structures composed primarily of coral rock formed from the calcium carbonate secreted by coral polyps. Many organisms contain very potent biochemical substances.

SEA ANEMONE *(Actinia)*
This solitary polyp has retractable, stinging tentacles used for defense and for the capture of prey. There are more than 800 species. Most live attached to rocky surfaces, although they can creep along the seabed.

CONSERVATION
Coral reefs can require many years to reach full size; they need at least 20 years to grow to the size of a soccer ball. Global warming is one of the major causes of the widespread deterioration of coral colonies. Other threats to coral reefs include tourism and pollution.

POWERFUL POISON
Despite its small size (less than 8 inches [20 cm]), this octopus is capable of killing a person. It releases extremely poisonous saliva into the water to stun and catch its prey, or it can inject the venom with a fatal bite. When it is disturbed, the iridescence of the rings on its skin glow more brightly. It is the most dangerous cephalopod.

CONTINUOUS BARRIER
This barrier reef is separated from the coast by saltwater lagoons. It is made up of colonies of coral polyps that secrete a hard exoskeleton of calcium carbonate. As the polyps divide and the colony grows, the reef becomes larger.

BLUE-RINGED OCTOPUS
(Hapalochlaena maculosa)

Types of Reefs

There are some 230,000 square miles (600,000 sq km) of coral reefs in the world's oceans. They were originally classified in 1842 by Charles Darwin, who described three principal types. Fringing, or coastal, reefs grow in shallow waters near the shoreline on the continental shelf. They support the most complex aquatic ecosystems. Barrier reefs, such as the Great Barrier Reef of Australia, are separated from continental landmasses by shallow saltwater lagoons that can become very wide and deep. Atolls, such as the Tahitian atoll, are ring-shaped. They surround a deep central lagoon, which is as rich in sealife as the coral reef itself. Atolls originate from fringing reefs that form around the sides of a volcanic island. Later, the volcano becomes submerged, either as a result of rising sea levels or because of the sinking of the volcanic island itself. The reefs continue to grow until only the ring of coral remains. Atolls are generally located far from continental landmasses.

Life in the Coral Reef

In addition to coral polyps, coral reefs house a large number of multicolored species (such as fish, tortoises, starfish, giant clams, snails, octopuses, sponges, tubular worms, sea urchins, anemones, and others). The inhabitants build a complex web of associations. Many animals can live together without competing for the same prey. The reef has countless small nooks that predators and prey use for shelter. Every food source in the reef is recycled through food chains. Food chains begin with diminutive single-celled algae that live inside the coral polyps. Algae also help to produce the limestone necessary for the construction of the reef.

BARRIER REEF
of Bora Bora in the Society Islands of French Polynesia.

SYMBIOSIS

This is a relationship between different organisms that live together in close association. In particular, mutualism is a type of symbiosis in which both organisms benefit. Although it is found in all habitats, symbiosis is especially prevalent in coastal areas of the ocean. A common example found in coral reefs is the clown fish, which lives among the tentacles of large sea anemones. It is protected from the anemone's stinging cells by a coating of mucus. While the anemone's stinging tentacles protect the fish from predators, the fish cleans the anemone by ingesting harmful material that collects on the anemone's outer covering.

LETHAL SPINE
This voracious predator attacks its prey by inflicting a painful wound with the venomous spines of its pectoral fins. It feeds on small shrimp and crabs as well as very large fish.

WHAT IS CORAL?

A piece of coral is the skeletal remains of microscopic animals. These animals are tubular polyps less than 0.3 inches (5 mm) in diameter. They affix themselves to a substrate by means of a hard basal disk. Each polyp has an opening that is connected to a gastric cavity that secretes the calcium carbonate substance that forms the reef. Coral can take many shapes; it can resemble such objects as trees, mushrooms, and flowers.

LIONFISH
(Pterois volitans)

CLOWN FISH
(Amphiprion ocellaris)

BARREL SPONGE
(Xestospongia testudinaria)

YELLOW-TUBED SPONGE
(Aplysina fistularis)

BLUE TUBE SPONGE
(Kallypilidion fascigera)

The Greenhouse Effect

Some gases in the atmosphere have the ability to trap the heat that arrives on Earth's surface from the Sun. All these gases work together to produce the greenhouse effect. Just mentioning the words, however, can trigger anxiety, since many believe that it is the principal cause of global warming. Its villainous reputation, however, sometimes obscures the fact that without the greenhouse effect there would be no life on Earth; the world would become frozen and lifeless without this phenomenon.

A Surprising Gas Ceiling

The Earth receives heat from the Sun. A portion of this heat is reflected from the Earth's surface back to space. The greenhouse gases in the atmosphere, however, trap a portion of the heat and return it back to Earth. In the process, these gases help to warm the surface and the lowest layer of the atmosphere (the troposphere).

-7.6⁰ F
(-22⁰ C)

This would be the average temperature of the Earth if there were no greenhouse effect, and the temperature would fluctuate greatly between day and night.

1 The Sun's rays pass through the atmosphere and reach the Earth's surface largely unimpeded.

ALBEDO
About 30% of the solar radiation that reaches the Earth is automatically reflected back into space owing to the planet's albedo—that is, the ratio between incident and reflected radiation.

ATMOSPHERE

GREENHOUSE GASES

3 Most of the radiation reflected by the Earth is radiated back toward the Earth's surface by the greenhouse gases. This process further heats the Earth and atmosphere.

2 Clouds, water, and soil absorb part of the incoming solar radiation and are thus warmed. Another portion of this radiation is reradiated back toward space as heat (infrared radiation).

4 Infrared radiation (heat) passes back and forth, heating the Earth's surface, although heating becomes less efficient each time.

SURFACE OF THE EARTH

The Carbon Cycle

Because carbon combines with oxygen to form carbon dioxide, the main greenhouse gas, scientists place special emphasis on observing the way carbon moves through nature. Carbon is a basic constituent of living things and is in continual movement through the biosphere.

385 ppm

The concentration of CO_2 in the atmosphere in 2008, a value that has not been observed in the past 420,000 years. Some authorities suspect that it has been about 20 million years since the concentrations have been this high.

Atmosphere: 750

92

121

Exchange between the ground and the air

The growth and death of plants and soil respiration

Soil: 1,580

The illustration shows the approximate quantity of carbon involved in the carbon cycle, measured in millions of tons.

Exchange between the ocean and atmosphere

The production of fossil fuels: 4,000

Fossil-fuel emissions: 5.5

Land plants: 540 to 610

60

Fires

Surface waters: 1,020

90

0.5

Changes in land use

1.5

Marine organisms: 3

Exchange between surface water and deep water

92

Petroleum and gas deposits: 300

Intermediate and deep waters: 28,000 to 40,000.

100

Coal deposits: 3,000

Surface sediments: 150

Dissolved organic carbon: 700

Marine rocks and sediments: 66,000,000 to 100,000,000

WHAT HUMANS CONTRIBUTE

Today, the atmosphere contains a high concentration of greenhouse gases. It is this increase in concentration that is believed to contribute to climate change. Much of the added amount is related to human activity. Specific greenhouse gases are described below.

GREENHOUSE GASES

Carbon dioxide makes up one-half of all the greenhouse gases in the atmosphere, followed by methane, nitrous oxides, and chlorofluorocarbons (CFCs). The percentage of the total amount of greenhouse gases in the atmosphere:

CARBON DIOXIDE (CO$_2$)
It is produced naturally through biological processes such as decomposition and combustion. In the past 250 years, however, human activities—in particular industrial processes, deforestation, and the use of vehicles powered by fossil fuels—have increased the levels of carbon dioxide.

METHANE (CH$_4$)
The simplest hydrocarbon. It is produced naturally during anaerobic decomposition—in others words, bacterial decomposition in which oxygen is not used.

CHLOROFLUOROCARBONS (CFCS)
These compounds were synthesized by human beings and are used in industry, especially for refrigeration. Although these compounds are nontoxic to human beings, they are very damaging to the ozone layer that protects the Earth from harmful solar radiation.

STRATOSPHERIC OZONE (O$_3$)
The ozone in the stratosphere provides protection from the Sun. The ozone at or near the Earth's surface (low-level ozone), which is produced in industrial processes and by the burning of fossil fuels, is an air pollutant and acts as a greenhouse gas.

NITROGEN OXIDES (NO$_x$)
These gases are also produced by industrial process and the burning of fossil fuels.

Climate Change

There is no longer room for doubt: average global temperatures are rising year after year, and the consequences of this change are beginning to manifest themselves. In addition, the concentration of greenhouse gases in the atmosphere is reaching levels unseen in many thousands of years, thanks to human activities. Are humans to blame for today's climate change? Are human activities only contributing factors, or do they have no effect at all? Knowing the answers to these questions is instrumental in deciding which steps to take to mitigate the effects of global warming, which could be one of the most dramatic events in the history of humanity.

Why Is the Planet Warming?

Despite the large amount of evidence showing that human activities definitively influence the composition of the atmosphere, it is not so clear the extent to which they contribute to warming, or whether they have even been the direct cause. In any case, there are other factors that need to be taken into account.

Earth's orbit fluctuates over long periods of time, and this has significant consequences, such as the periods of glaciation (the Ice Ages) that the planet has experienced. The effect this fluctuation might have on present climate change is not known, however.

-0.9° F
(0.5° C)

This is the increase in average global temperature since people began to measure it, less than two centuries ago.

SOLAR ACTIVITY
The Sun is a star that goes through periods of greater and lesser activity—a cycle that also has an enormous effect on climate. It is not clear, however, what effects this activity might have on present climate change.

— Annual irradiation
— Sunspots
— Solar radiation index

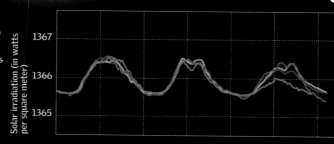

Solar irradiation (in watts per square meter)

1367

1366

1365

120,000 years

The time since the last major glacial period, which ended only 10,000 years ago. Some researchers believe that the current warming climate is associated with the transition to warmer times from the last Ice Age.

ALBEDO
Ice reflects most of the light and energy that it receives from the Sun back into space. When areas of ice shrink, the albedo decreases, and the planet absorbs more energy and becomes warmer.

Lines of magnetic force

Magnetic North Pole

Geographic North Pole

EARTH'S MAGNETIC FIELD
The Earth's magnetic field is constantly changing. In the past, the magnetic poles of the field have reversed, and at times, they have even been located over the Equator. This variation affects the climate indirectly since it affects the way in which electromagnetic particles from the Sun, called the solar wind, reach the Earth. Its relationship to present climate change is not clear.

Magnetic South Pole

Geographic South Pole

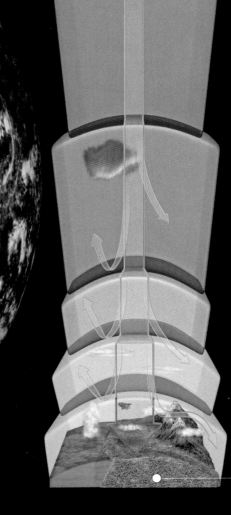

GREENHOUSE GASES
They are essential for maintaining life on Earth. An increase in the concentration of these gases in the atmosphere, however, could be the reason for the rise in average global temperature. Human activities have raised the carbon dioxide concentrations to their highest level in many thousands, perhaps many millions, of years.

INDUSTRIAL EMISSIONS
A large part of the energy that drives industry comes from the burning of fossil fuels, which produces enormous amounts of greenhouse gases that are released into the atmosphere.

DEFORESTATION
As overall biomass is reduced, the environment's capacity to absorb carbon dioxide in huge amounts is diminished; this leaves a higher concentration of the greenhouse gases in the atmosphere.

TRANSPORT
At present, the machines of the world are powered by petroleum-derived fuels, which constitute one of the most important sources of atmospheric carbon dioxide. The substitution of biofuels will not affect these emission levels.

The study of fossils provides information about the Earth's climate at different times in the past.

The Ozone Hole

Earth is protected in large measure from harmful solar radiation by an invisible layer of gas containing ozone molecules. Each spring, however, the concentration of ozone abruptly diminishes over the polar regions, especially over Antarctica. Although this phenomenon was initially considered to be part of a natural cycle, scientists became alarmed when they discovered that synthetic gases could be responsible for causing the "ozone hole" to deepen in a worrisome fashion during the past decades.

A Stable Protective Shield

The ozone layer that protects the Earth from ultraviolet B radiation lies at an altitude that varies between 6 and 30 miles (10–50 km) above the Earth's surface. Ozone is also found near the Earth's surface. Low-level ozone is a product of pollution and can be harmful to plants and animals.

Sun

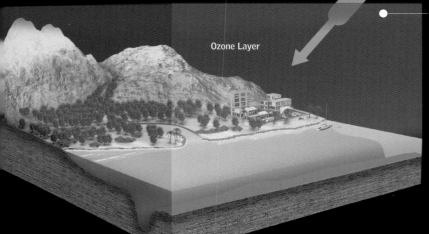

Ozone Layer

UVB Filter
The ozone filters most of the ultraviolet B radiation (UVB) from the Sun and converts the radiation into heat. Unfiltered, this type of radiation can kill microorganisms, damage plants and animals, and cause cancer in humans.

Endless Cycle
When radiation from the Sun strikes a molecule of ozone, the molecule breaks apart, producing highly reactive oxygen. The ozone molecule then reforms, releasing heat in the process.

Trouble in Springtime

Every spring, year after year, the concentration of atmospheric ozone over Antarctica falls sharply, allowing a greater amount of UVB to pass through. The ozone layer is restored in the summer.

Size in millions of square kilometers (10 million square km = 3.9 million square miles)

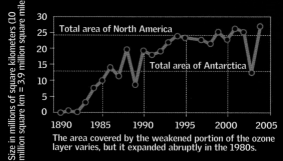

Total area of North America

Total area of Antarctica

30
25
20
15
10
5
0

1890   1985   1990   1995   2000   2005

The area covered by the weakened portion of the ozone layer varies, but it expanded abruptly in the 1980s.

VARIATION OF THE OZONE HOLE
This series of images shows the measurements of the ozone hole over Antarctica in the month of September.

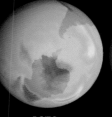

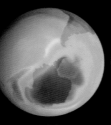

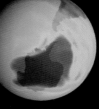

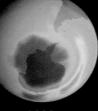

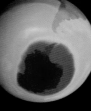

1979   1982   1985   1988   1994

30,000

This is the number of molecules of ozone in the atmosphere that can be destroyed by each chlorine atom.

▢	350 du
▢	320 du
▢	285 du
▢	220 du (hole)

Lethal Attack

Although it was once believed that the weakening of the ozone layer was the result of natural causes, it was soon discovered that the emission of certain manufactured gases can be highly destructive to it, although it is not known precisely to what degree.

CHLOROFLUOROCARBONS (CFCs)

First created in the 1930s, CFCs are derived from hydrocarbons in which hydrogen atoms are replaced by atoms of fluorine and chlorine. For many years they were ideal for use as refrigerants, fire-extinguishing agents, and aerosol propellants because of their low toxicity and their physical and chemical stability. It was subsequently discovered, however, that they are very destructive to the ozone layer.

SOURCES OF ATMOSPHERIC CHLORINE

Natural processes **18%**

Human activities **82%**

A Ray of Hope

Alarmed by the rapid fall in levels of ozone in the ozone layer, 191 countries signed the 1987 Montreal Protocol, which obligated the signatory countries to reduce their emissions of gases that affect the ozone layer. The protocol is considered to be the first global success in the fight to protect the environment.

From 60° N to 60° S

Change in chlorine in the atmosphere since the Montreal Protocol

The first stage in the recovery of the ozone hole (1997)

1980 1985 1990 1995 2000 2005

THE PROCESS

1. The molecules of ozone and CFC exist together high in the atmosphere.

Ozone — CFC

2. The UV radiation breaks apart the CFC molecule, leaving free atoms of chlorine.

UV

3. The chlorine atom is highly reactive, and it breaks apart the ozone molecule to combine with an atom of oxygen.

Chlorine monoxide

Ozone = Oxygen

4. A free oxygen atom in the atmosphere is also highly reactive, and it breaks apart the ClO molecule, once again freeing the chlorine atom.

Oxygen

5. The free chlorine attacks a new molecule of ozone, repeating the process.

Chlorine monoxide

Ozone = Oxygen

0.1 inch (3 mm)

What the thickness of the ozone layer around the world would be if the gas were isolated under ideal conditions of pressure and temperature.

1998 2000 2001 2007 2010 2015

LIGHTNING

The product of enormous electrical charges that build up inside clouds, it is followed by a great rumble—thunder—caused by the release of sound energy.

CHAPTER 4

WEATHER AND CLIMATOLOGY

The Earth's climate system is a large machine powered by the Sun and by five contributing elements: the atmosphere, biosphere, hydrosphere, cryosphere, and lithosphere. The interaction between these subsystems determines the climate in every zone of the planet as well as the temperature, rainfall levels, winds, etc.

Global Equilibrium

The Sun's radiation delivers a large amount of energy, which propels the Earth's extraordinary mechanism called the climatic system. The components of this complex system are the atmosphere, hydrosphere, lithosphere, cryosphere, and biosphere. All these components are constantly interacting with one another via an interchange of materials and energy. Weather and climatic phenomena of the past—as well as of the present and the future—are the combined expression of Earth's climatic system.

WINDS

The atmosphere is always in motion. Heat displaces masses of air, and this leads to the general circulation of the atmosphere.

Atmosphere

Part of the energy received from the Sun is captured by the atmosphere. The other part is absorbed by the Earth or reflected in the form of heat. Greenhouse gases heat up the atmosphere by slowing the release of heat to space.

Biosphere

Living beings (such as plants) influence weather and climate. They form the foundations of ecosystems, which use minerals, water, and other chemical compounds. They contribute materials to other subsystems.

PRECIPITATION

Water condensing in the atmosphere forms droplets, and gravitational action causes them to fall on different parts of the Earth's surface.

EVAPORATION

The surfaces of water bodies maintain the quantity of water vapor in the atmosphere within normal limits.

about 10%

Albedo of the tropical forests.

HEAT

Night and day, coastal breezes exchange energy between the hydrosphere and the lithosphere.

MARINE CURRENTS

Hydrosphere

The hydrosphere is the name for all water in liquid form that is part of the climatic system. Most of the lithosphere is covered by liquid water, and some of the water even circulates through it.

3%

Albedo of the bodies of water.

SOLAR RADIATION

About 50 percent of the solar energy reaches the surface of the Earth, and some of this energy is transferred directly to different layers of the atmosphere. Much of the available solar radiation leaves the air and circulates within the other subsystems. Some of this energy escapes to outer space.

ALBEDO

The percentage of solar radiation reflected by the climatic subsystems.

Sun

Essential for climatic activity. The subsystems absorb, exchange, and reflect energy that reaches the Earth's surface. For example, the biosphere incorporates solar energy via photosynthesis and intensifies the activity of the hydrosphere.

Cryosphere

Represents regions of the Earth covered by ice. Permafrost exists where the temperature of the soil or rocks is below zero. These regions reflect almost all the light they receive and play a role in the circulation of the ocean, regulating its temperature and salinity.

Lithosphere

This is the uppermost solid layer of the Earth's surface. Its continual formation and destruction change the surface of the Earth and can have a large impact on weather and climate. For example, a mountain range can act as a geographic barrier to wind and moisture.

50%

The albedo of light clouds.

SUN

80%

Albedo of recently fallen snow.

HEAT

HUMAN ACTIVITY

SMOKE
Particles that escape into the atmosphere can retain their heat and act as condensation nuclei for precipitation.

RETURN TO THE SEA
UNDERGROUND CIRCULATION
The circulation of water is produced by gravity. Water from the hydrosphere infiltrates the lithosphere and circulates therein until it reaches the large water reservoirs of lakes, rivers, and oceans.

ASHES
Volcanic eruptions bring nutrients to the climatic system where the ashes fertilize the soil. Eruptions also block the rays of the Sun and thus reduce the amount of solar radiation received by the Earth's surface. This causes cooling of the atmosphere.

GREENHOUSE EFFECT
Some gases in the atmosphere are very effective at retaining heat. The layer of air near the Earth's surface acts as a shield that establishes a range of temperatures on it, within which life can exist.

SOLAR ENERGY

OZONE LAYER

ATMOSPHERE

STRATOSPHERE

TROPOSPHERE

TROPOPAUSE

STRATOPAUSE

Climate Zones

Different places in the world, even if far removed from each other, can be grouped into climate zones—that is, into regions that are homogeneous relative to climatic elements, such as temperature, pressure, rain, and humidity. There is some disagreement among climatologists about the number and description of each of these regions, but the illustrations given on this map are generally accepted.

Ice cap

PLAINS AND URBANIZATION

Human settlements

Fertile soil, stable climate

Fruit trees

Natural brush

Agriculture

Hudson Bay

NORTH AMERICA

Rocky Mountains

Appalachian Mountains

TEMPERATE
Characterized by pleasant temperatures and moderate rains throughout the year. Winters are mild, with long, frost-free periods. Temperate regions are ideal for most agricultural products.

HOUSTON, U.S.
Annual precipitation of 46 inches (1,170 mm)

mm
1,000
500
250
0

°C
40
20
0
-20

J F M A M J J A S O N D

CENTRAL AMERICA

Pacific Ocean

59⁰ F
(15⁰ C)

is the average annual temperature of the Earth.

Atlantic Ocean

TROPICAL
High temperatures throughout the year, combined with heavy rains, are typical for this climate. About half of the world's population lives in regions with a tropical climate. Vegetation is abundant, and humidity is high because the water vapor in the air is not readily absorbed.

RAINFOREST OR JUNGLE

Tropical fruits and flowers

Plentiful water sources

Green and fertile soil

Layers of vegetation

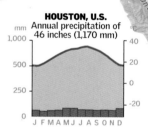

Cordillera de los Andes

Amazon basin

SOUTH AMERICA

Pampas region

Patagonia

DESERT
Intermittent water

Sea of dunes

Sparse vegetation

MANAUS, BRAZIL
Annual precipitation 75 inches (1,900 mm)

mm
1,000
500
250
0

°C
40
20
0
-20

J F M A M J J A S O N D

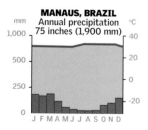

Temperature and Rains

The temperature of the Earth depends on the energy from the Sun, which is not distributed equally at all latitudes. Only 5 percent of sunlight reaches the surface at the poles, whereas this figure rises to 75 percent at the Equator. Rain is an atmospheric phenomenon. Clouds contain millions of drops of water, which collide to form larger drops. The size of the drops increases until they are too heavy to be supported by air currents, and they fall as rain.

DRY
Lack of rain controls the arid climate in desert or semidesert regions, the result of the atmospheric circulation of air. In these regions, dry air descends, leaving the sky clear, with many hours of burning Sun.

FORESTS AND LAKES

Coniferous forest

Deciduous trees

Juniper brush

Lakes

MOSCOW, RUSSIA
Annual precipitation 25 inches (624 mm)

mm
1,000
500
250
0

°C
40
20
0
-20

J F M A M J J A S O N D

COLD

Very cold winters, with frequent freezing at night, are typical of these regions. In these zones, the climate changes more often than anywhere else. In most cold climate regions, the landscape is covered by natural vegetation.

Siberia

Plains of Siberia

ASIA

East European Plain

EUROPE

Alps

Black Sea

Caspian Sea

Sahara

Arabian Peninsula

Himalayas

Congo basin

Indian Ocean

AFRICA

POLAR MOUNTAINOUS CLIMATE

Mountains create their own climate that is somewhat independent of their location. Near the poles, the polar climate is dominated by very low temperatures, strong and irregular winds, and almost perpetual snow. The mountain peaks lack vegetation.

Eternal snow on the mountains

TUNDRA AND TAIGA

Sparse conifers

Lichens

12° F (6.5° C)

is the temperature decrease for every 3,300 feet (1,000 m) of increase in elevation.

OCEANIA

Gibson Desert

LHASA, TIBET
Annual precipitation 16 inches (408 mm)

mm
1,000
500
250
0

°C
40
20
0
-20

J F M A M J J A S O N D

Dry soil

Sand

TIMBUKTU, MALI
Annual precipitation 9 inches (232 mm)

mm
1,000
500
250
0

°C
40
20
0
-20

J F M A M J J A S O N D

Köppen Climate Classification

In 1936 Russian-born climatologist Wladimir Köppen presented a climatological classification based on temperature and precipitation. The table provides a broad overview of the approximate distribution of climates on the terrestrial globe. Köppen classification does not discuss climatic regions but rather the type of climate found in a given location according to specific parameters.

Latitudes
80°
60°
40°
20°
0°
20°
40°
60°

KEY

- Tropical forests, without a dry season
- Tropical savanna, with a dry winter
- Steppes (semiarid)
- Desert (arid)
- Temperate humid, without a dry season
- Temperate, with a dry winter
- Temperate, with a dry summer
- Tundra
- Glacial
- Mountain climate
- Temperate cold continental (hot summer)
- Temperate cold continental (cold summer)
- Temperate cold continental (subarctic)

Atmospheric Dynamics

The atmosphere is a dynamic system. Temperature changes and the Earth's motion are responsible for horizontal and vertical air displacement. Here the air of the atmosphere circulates between the poles and the Equator in horizontal bands within different latitudes. Moreover, the characteristics of the Earth's surface alter the path of the moving air, causing zones of differing air densities. The relations that arise among these processes influence the climatic conditions of our planet.

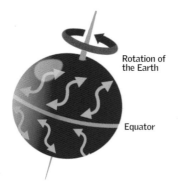

Rotation of the Earth

Equator

CORIOLIS FORCE

The Coriolis effect is an apparent deflection of the path of an object that moves within a rotating coordinate system. The Coriolis effect appears to deflect the trajectory of the winds that move over the surface of the Earth, because the Earth moves beneath the winds. This apparent deflection is to the right in the Northern Hemisphere and to the left in the Southern Hemisphere. The effect is only noticeable on a large scale because of the rotational velocity of the Earth.

FERREL CELL

A part of the air in the Hadley cells follows its course toward the poles to a latitude of 60° N and 60° S.

Intertropical Convergence Zone (ITCZ)

TRADE WINDS

These winds blow toward the Equator.

High and Low Pressure

Warm air rises and causes a low-pressure area (cyclone) to form beneath it. As the air cools and descends, it forms a high-pressure area (anticyclone). Here the air moves from an anticyclonic toward a cyclonic area as wind. The warm air, as it is displaced and forced upward, leads to the formation of clouds.

6 The masses of cold air lose their mobility.

1 Masses of cold air descend and prevent clouds from forming.

5 The rising air leads to the formation of clouds.

3 The wind blows from a high-toward a low-pressure area.

A

B

2 The descending air forms an area of high pressure (anticyclone).

4 Warm air rises and forms an area of low pressure (cyclone).

− Low-pressure area

+ High-pressure area

Jet-stream currents

Changes in Circulation

Irregularities in the topography of the surface, abrupt changes in temperature, and the influence of ocean currents can alter the general circulation of the atmosphere. These circumstances can generate waves in the air currents that are, in general, linked to the cyclonic zones. It is in these zones that storms originate, and they are therefore studied with great interest. However, the anticyclone and the cyclone systems must be studied together because cyclones are fed by currents of air coming from anticyclones.

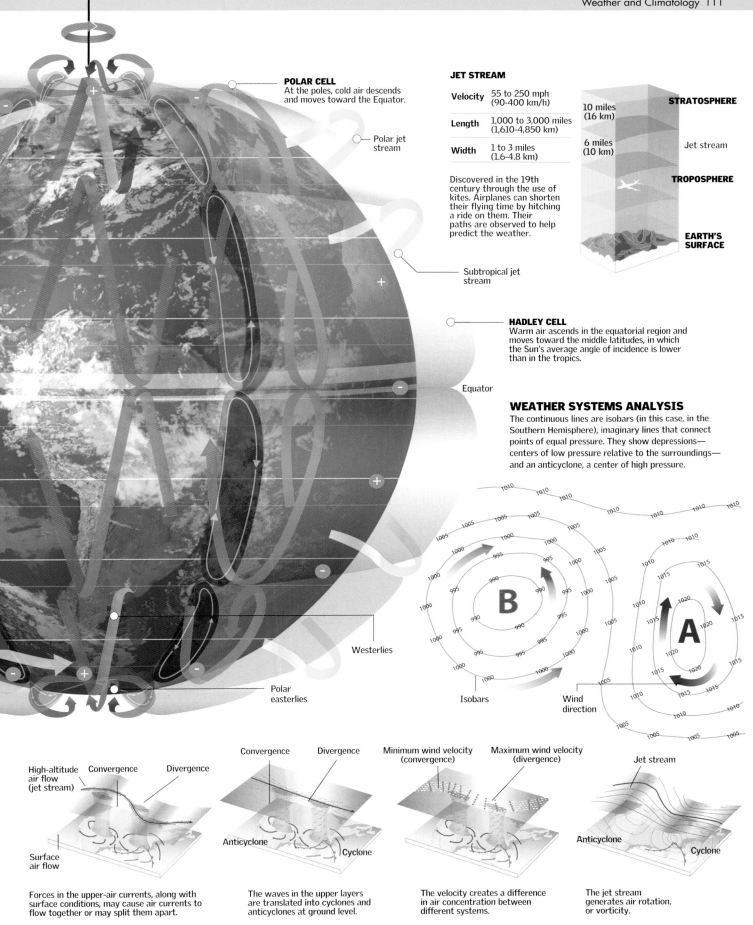

POLAR CELL
At the poles, cold air descends and moves toward the Equator.

Polar jet stream

JET STREAM

Velocity	55 to 250 mph (90-400 km/h)
Length	1,000 to 3,000 miles (1,610-4,850 km)
Width	1 to 3 miles (1.6-4.8 km)

Discovered in the 19th century through the use of kites. Airplanes can shorten their flying time by hitching a ride on them. Their paths are observed to help predict the weather.

10 miles (16 km)

STRATOSPHERE

6 miles (10 km)

Jet stream

TROPOSPHERE

EARTH'S SURFACE

Subtropical jet stream

HADLEY CELL
Warm air ascends in the equatorial region and moves toward the middle latitudes, in which the Sun's average angle of incidence is lower than in the tropics.

Equator

WEATHER SYSTEMS ANALYSIS
The continuous lines are isobars (in this case, in the Southern Hemisphere), imaginary lines that connect points of equal pressure. They show depressions—centers of low pressure relative to the surroundings—and an anticyclone, a center of high pressure.

Westerlies

Polar easterlies

B

A

Isobars

Wind direction

High-altitude air flow (jet stream)

Convergence   Divergence

Surface air flow

Forces in the upper-air currents, along with surface conditions, may cause air currents to flow together or may split them apart.

Convergence   Divergence

Anticyclone   Cyclone

The waves in the upper layers are translated into cyclones and anticyclones at ground level.

Minimum wind velocity (convergence)   Maximum wind velocity (divergence)

The velocity creates a difference in air concentration between different systems.

Jet stream

Anticyclone   Cyclone

The jet stream generates air rotation, or vorticity.

Collision

When two air masses with different temperatures and moisture content collide, they cause atmospheric disturbances. When the warm air rises, its cooling causes water vapor to condense and the formation of clouds and precipitation. A mass of warm and light air is always forced upward, while the colder and heavier air acts like a wedge. This cold-air wedge undercuts the warmer air mass and forces it to rise more rapidly. This effect can cause variable, sometimes stormy, weather.

Cold Fronts

These fronts occur when cold air is moved by the wind and collides with warmer air. Warm air is driven upward. The water vapor contained in the air forms cumulus clouds, which are rising, dense white clouds. Cold fronts can cause the temperature to drop by 10° to 30° F (about 5°-15° C) and are characterized by violent and irregular winds. Their collision with the mass of ascending water vapor will generate rain, snow flurries, and snow. If the condensation is rapid, heavy downpours, snowstorms (during the cold months), and hail may result. In weather maps, the symbol for a cold front is a blue line of triangles indicating the direction of motion.

Very dense clouds that rise to a considerable altitude

Cold front

Warm front

Cold air

Warm air

Cool air

Severe imbalance in the cold front

The cold front forces the warm air upward, causing storms.

Behind the cold front, the sky clears and the temperature drops.

There could be precipitation in the area with warm weather.

Rossby Waves

Large horizontal atmospheric waves that are associated with the polar-front jet stream. They may appear as large undulations in the path of the jet stream. The dynamics of the climatic system are affected by these waves because they promote the exchange of energy between the low and high latitudes and can even cause cyclones to form.

1 A long Rossby wave develops in the jet stream of the high troposphere.

2 The Coriolis effect accentuates the wave action in the polar air current.

3 The formation of a meander of warm and cold air can provide the conditions needed to generate cyclones.

STATIONARY FRONTS

These fronts occur when there is no forward motion of warm or cold air—that is, both masses of air are stationary. This type of condition can last many days and produces only altocumulus clouds. The temperature also remains stable, and there is no wind except for some flow of air parallel to the line of the front. There could be some light precipitation.

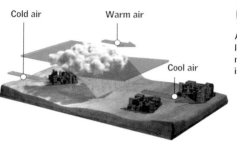

Cold air   Warm air   Cool air

OCCLUDED FRONTS

When the cold air replaces the cool air at the surface, with a warm air mass above, a cold occlusion is formed. A warm occlusion occurs when the cool air rises above the cold air. These fronts are associated with rain or snow, cumulus clouds, slight temperature fluctuations, and light winds.

Entire Continents

Fronts stretch over large geographic areas. In this case, a cold front causes storm perturbations in western Europe. But to the east, a warm front, extending over a wide area of Poland, brings light rain. These fronts can gain or lose force as they move over the Earth's surface depending on the global pressure system.

GERMANY   POLAND   BELARUS   UKRAINE   FRANCE

125 miles
(200 km)

A warm front can be 125 miles (200 km) long. A cold front usually covers about 60 miles (100 km). In both cases, the altitude is roughly 0.6 mile (1 km).

Cold front   Warm air   Cold air   Cool air

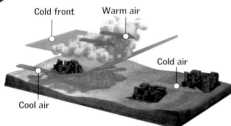

KEY

Surface cold front   Surface warm front

Warm Fronts

These are formed by the action of winds. A mass of warm air occupies a place formerly occupied by a mass of cold air. The speed of the cold air mass, which is heavier, decreases at ground level by friction, through contact with the ground. The warm front ascends and slides above the cold mass. This typically causes precipitation at ground level. Light rain, snow, or sleet are typically produced, with relatively light winds. The first indications of warm fronts are cirrus clouds, some 600 miles (1,000 km) in front of the advancing low pressure center. Next, layers of stratified clouds, such as the cirrostratus, altostratus, and nimbostratus, are formed while the pressure is decreasing.

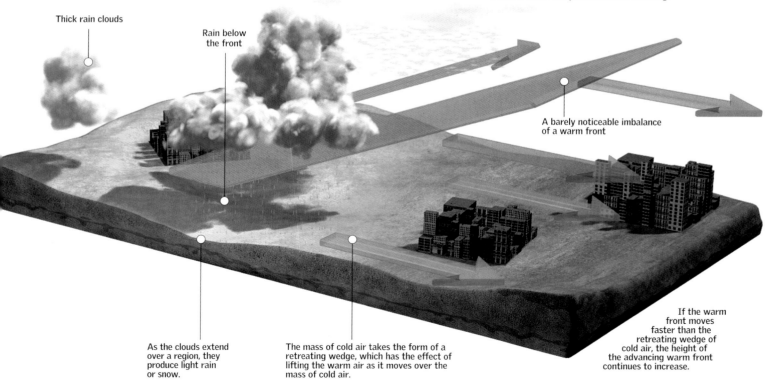

Thick rain clouds

Rain below the front

A barely noticeable imbalance of a warm front

As the clouds extend over a region, they produce light rain or snow.

The mass of cold air takes the form of a retreating wedge, which has the effect of lifting the warm air as it moves over the mass of cold air.

If the warm front moves faster than the retreating wedge of cold air, the height of the advancing warm front continues to increase.

Monsoons

The strong humid winds that usually affect the tropical zone are called monsoons, an Arabic word meaning "seasonal winds." During summer in the Northern Hemisphere, they blow across Southeast Asia, especially the Indian peninsula. Conditions change in the winter, and the winds reverse and shift toward the northern regions of Australia. This phenomenon, which is also frequent in continental areas of the United States, is part of an annual cycle that, as a result of its intensity and its consequences, affects the lives of many people.

AREAS AFFECTED BY MONSOONS

This phenomenon affects the climates in low latitudes, from West Africa to the western Pacific. In the summer, the monsoon causes the rains in the Amazon region and in northern Argentina. There in the winter rain is usually scarce.

Predominant direction of the winds during the month of July

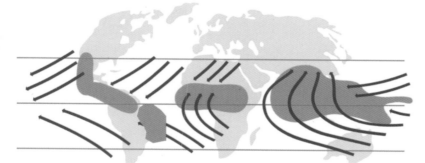

THE MONSOON OF NORTH AMERICA

Pre-monsoon. Month of May.

Monsoon. Month of July.

Cross section (enlarged area)

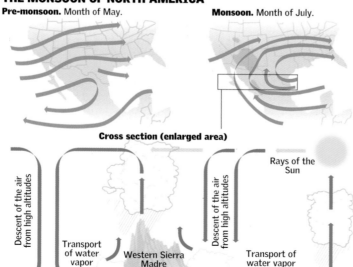

Descent of the air from high altitudes

Transport of water vapor

Western Sierra Madre

Descent of the air from high altitudes

Rays of the Sun

Transport of water vapor

Pacific Ocean

Gulf of California

Gulf of Mexico

HOW MONSOONS ARE CREATED IN INDIA

End of the monsoon	Beginning of the monsoon	Cold and dry winds	Cold and humid winds	Cyclone (low pressure)	Anticyclone (high pressure)

1 THE CONTINENT COOLS

After the summer monsoon, the rains stop and temperatures in Central and South Asia begin to drop. Winter begins in the Northern Hemisphere.

Northern Hemisphere
It is winter. The rays of the Sun are oblique, traveling a longer distance through the atmosphere to reach the Earth's surface. Thus they are spread over a larger surface, so the average temperature is lower than in the Southern Hemisphere.

Rays of the Sun

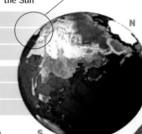

Southern Hemisphere
It is summer. The rays of the Sun strike the surface at a right angle; they are concentrated in a smaller area, so the temperature on average is higher than in the Northern Hemisphere.

2 FROM THE CONTINENT TO THE OCEAN

The masses of cold and dry air that predominate on the continent are displaced toward the ocean, whose waters are relatively warmer.

Arabian Sea

3 OCEAN STORMS

A cyclone located in the ocean draws the cold winds from the continent and lifts the somewhat warmer and more humid air, which returns toward the continent via the upper layers of the atmosphere.

INTERTROPICAL INFLUENCE

The circulation of the atmosphere between the tropics influences the formation of monsoon winds. The trade winds that blow toward the Equator from the subtropical zones are pushed by the Hadley cells and deflected in their course by the Coriolis effect. Winds in the tropics occur within a band of low pressure around the Earth called the Intertropical Convergence Zone (ITCZ). When this zone is seasonally displaced in the warm months of the Northern Hemisphere toward the north, a summer monsoon occurs.

Limit of the Intertropical Convergence Zone (ITCZ)

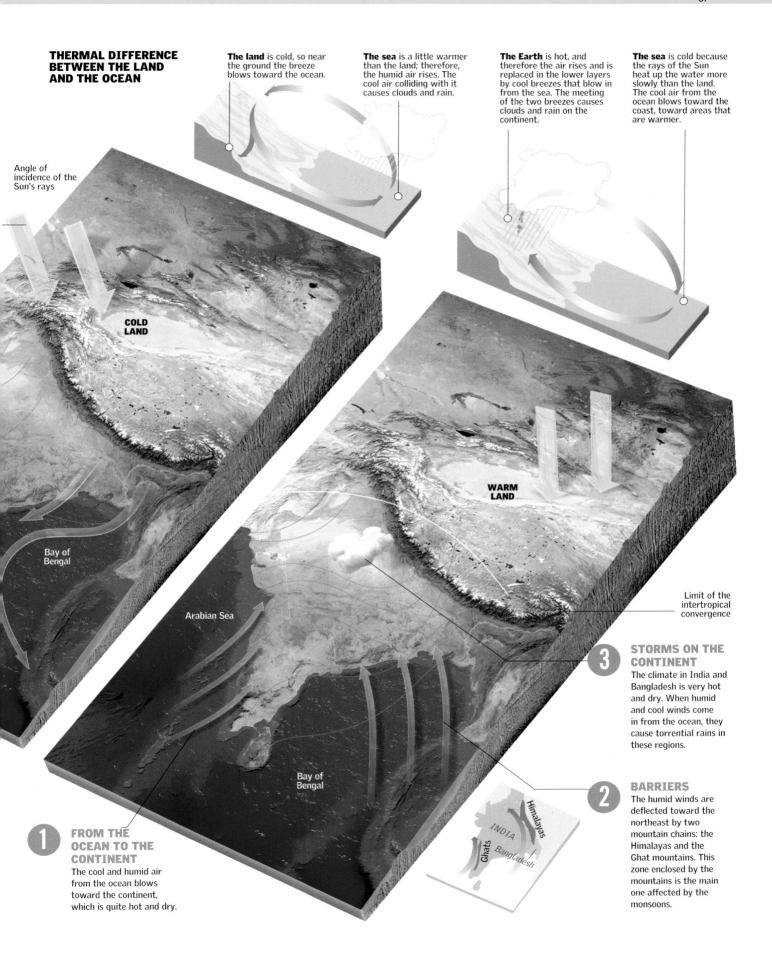

THERMAL DIFFERENCE BETWEEN THE LAND AND THE OCEAN

The land is cold, so near the ground the breeze blows toward the ocean.

The sea is a little warmer than the land; therefore, the humid air rises. The cool air colliding with it causes clouds and rain.

The Earth is hot, and therefore the air rises and is replaced in the lower layers by cool breezes that blow in from the sea. The meeting of the two breezes causes clouds and rain on the continent.

The sea is cold because the rays of the Sun heat up the water more slowly than the land. The cool air from the ocean blows toward the coast, toward areas that are warmer.

Angle of incidence of the Sun's rays

COLD LAND

WARM LAND

Bay of Bengal

Arabian Sea

Bay of Bengal

Limit of the intertropical convergence

3 STORMS ON THE CONTINENT
The climate in India and Bangladesh is very hot and dry. When humid and cool winds come in from the ocean, they cause torrential rains in these regions.

2 BARRIERS
The humid winds are deflected toward the northeast by two mountain chains: the Himalayas and the Ghat mountains. This zone enclosed by the mountains is the main one affected by the monsoons.

Himalayas

INDIA

Bangladesh

Ghats

1 FROM THE OCEAN TO THE CONTINENT
The cool and humid air from the ocean blows toward the continent, which is quite hot and dry.

Capricious Forms

Clouds are masses of large drops of water and ice crystals. They form because the water vapor contained in the air condenses or freezes as it rises through the troposphere. How the clouds develop depends on the altitude and the velocity of the rising air. Cloud shapes are divided into three basic types: cirrus, cumulus, and stratus. They are also classified as high, medium, and low depending on the altitude they reach above sea level. They are of meteorological interest because they indicate the behavior of the atmosphere.

TYPES OF CLOUDS

NAME	MEANING
CIRRUS	FILAMENT
CUMULUS	AGGLOMERATION
STRATUS	BLANKET
NIMBUS	RAIN

Troposphere

The layer closest to the Earth and in which meteorological phenomena occur, including the formation of clouds.

Exosphere

300 miles
(500 km)

Mesosphere

50 miles
(90 km)

Stratosphere

30 miles
(50 km)

Troposphere

6 miles
(10 km)

0

HOW THEY ARE FORMED

Clouds are formed when the rising air cools to the point where it cannot hold the water vapor it contains. In such a circumstance, the air is said to be saturated, and the excess water vapor condenses. Cumulonimbus clouds are storm clouds that can reach a height of 43,000 feet (13,000 m) and contain more than 150,000 tons of water.

CONVECTION
The heat of the Sun warms the air near the ground, and because it is less dense than the surrounding air, it rises.

CONVERGENCE
When the air coming from one direction meets air from another direction, it is pushed upward.

GEOGRAPHIC ELEVATION
When the air encounters mountains, it is forced to rise. This phenomenon explains why there are often clouds and rain over mountain peaks.

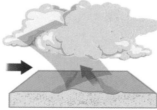

PRESENCE OF A FRONT
When two masses of air with different temperatures meet at a front, the warm air rises and clouds are formed.

TROPOS

HIGH CLOUDS

6 miles
(10km)

-67° F
(-55° C)
Temperature in the upper part of the troposphere.

CIRROSTRATUS
A very extensive cloud that eventually covers the whole sky and has the form of a transparent, fibrous-looking veil.

MEDIUM CLOUDS

2.5 miles
(4 km)

14° F
(-10° C)
The temperature of the middle part of the troposphere.

CUMULONIMBUS
A storm cloud. It portends intense precipitation in the form of rain, hail, or snow. Its color is white.

CUMULUS
A cloud that is generally dense with well-defined outlines. Cumulus clouds can resemble a mountain of cotton.

LOW CLOUDS

1.2 miles
(2 km)

50° F
(10° C)
Temperature of the lower part of the troposphere.

59° F
(15° C)
Temperature at the Earth's surface

0 miles (0 km)

HERE

CIRRUS
A high, thin cloud with white, delicate filaments composed of ice crystals.

CIRROCUMULUS
A cloud formation composed of very small, granulated elements spaced more or less regularly.

ALTOCUMULUS
A formation of rounded clouds in groups that can form straight or wavy rows.

1802

The year that British meteorologist Luke Howard carried out the first scientific study of clouds

ALTOSTRATUS
Large, nebulous, compact, uniform, slightly layered masses. Altostratus does not entirely block out the Sun. It is bluish or gray.

STRATOCUMULUS
A cloud that is horizontal and very long. It does not blot out the Sun and is white or gray in color.

NIMBOSTRATUS
Nimbostratus portends more or less continuous precipitation in the form of rain or snow that, in most cases, reaches the ground.

STRATUS
A low cloud that extends over a large area. It can cause drizzle or light snow. Stratus clouds can appear as a gray band along the horizon.

The Inside

The altitude at which clouds are formed depends on the stability of the air and the humidity. The highest and coldest clouds have ice crystals. The lowest and warmest clouds have drops of water. There are also mixed clouds. There are 10 classes of clouds depending on their height above sea level. The highest clouds begin at a height of 2.5 miles (4 km). The mid-level begins at a height of 1.2 to 2.5 miles (2-4 km) and the lowest at 1.2 miles (2 km) high.

1.2-5 mi
(2-8 km)
Thickness of a storm cloud

150,000
tons of water
can be contained in a storm cloud.

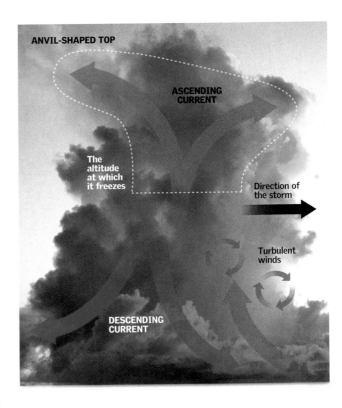

ANVIL-SHAPED TOP

ASCENDING CURRENT

The altitude at which it freezes

Direction of the storm

Turbulent winds

DESCENDING CURRENT

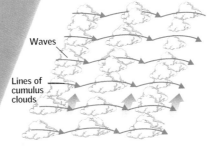

Waves

Lines of cumulus clouds

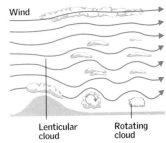

Wind

Lenticular cloud

Rotating cloud

SPECIAL FORMATIONS

CLOUD STREETS
The form of the clouds depends on the winds and the topography of the terrain beneath them. Light winds usually produce lines of cumulus clouds positioned as if along streets. Such waves can be created by differences in surface heating.

LENTICULAR CLOUDS
Mountains usually create waves in the atmosphere on their lee side, and on the crest of each wave lenticular clouds are formed that are held in place by the waves. Rotating clouds are formed by turbulence near the surface.

The Rain Announces It's Coming

The air inside a cloud is in continuous motion. This process causes the drops of water or the crystals of ice that constitute the cloud to collide and join together. In the process, the drops and crystals become too big to be supported by air currents and they fall to the ground as different kinds of precipitation. A drop of rain has a diameter 100 times greater than a droplet in a cloud. The type of precipitation depends on whether the cloud contains drops of water, ice crystals, or both. Depending on the type of cloud and the temperature, the precipitation can be liquid water (rain) or solid (snow or hail).

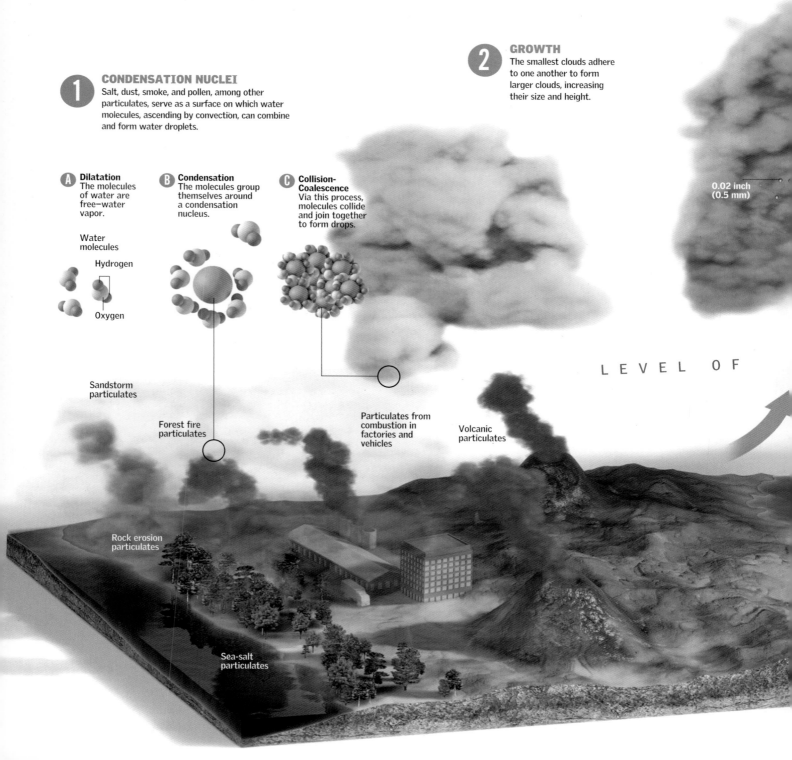

1 CONDENSATION NUCLEI
Salt, dust, smoke, and pollen, among other particulates, serve as a surface on which water molecules, ascending by convection, can combine and form water droplets.

2 GROWTH
The smallest clouds adhere to one another to form larger clouds, increasing their size and height.

A Dilatation
The molecules of water are free–water vapor.

Water molecules

Hydrogen

Oxygen

B Condensation
The molecules group themselves around a condensation nucleus.

C Collision-Coalescence
Via this process, molecules collide and join together to form drops.

0.02 inch (0.5 mm)

LEVEL OF

Sandstorm particulates

Forest fire particulates

Particulates from combustion in factories and vehicles

Volcanic particulates

Rock erosion particulates

Sea-salt particulates

3 MATURATION
Mature clouds have very strong ascending currents, leading to protuberances and rounded formations. Convection occurs.

• 4 miles (7 km)

-22° F
(-30° C)
When the air cools, it descends and is then heated again, repeating the cycle.

The air cools. The water vapor condenses and forms microdroplets of water.

• 0.6-1.2 miles
(1-2 km)

4 RAIN
The upper part of the cloud spreads out like an anvil, and the rain falls from the lower cloud, producing descending currents.

• 6 miles
(10 km)

ANVIL-SHAPED

STORM CLOUD

Coalescence
The microdroplets continue to collide and form bigger drops.

Heavier drops fall onto a lower cloud as fine rain.

— 0.04 inch
(1 mm)

5 DISSIPATION
The descending currents are stronger than the ascending ones and interrupt the feeding air, causing the cloud to disintegrate.

Low, thin clouds contain tiny droplets of water and therefore produce rain.

When they begin to fall, the drops have a size of 0.02 inch (0.5 mm), which is reduced as they fall since they break apart.

0.2 inch
(5 mm)

0.04 inch
(1 mm)

CONDENSATION

68° F
(20° C)
The hot air rises.

• 0 miles
(0 km)

— 0.07 inch
(2 mm)

26,875 trillion molecules occupy 1 cubic millimeter under normal atmospheric conditions.

Lost in the Fog

When atmospheric water vapor condenses near the ground, it forms fog and mist. The fog consists of small droplets of water mixed with smoke and dust particles. Physically the fog is a cloud, but the difference between the two lies in their formation. A cloud develops when the air rises and cools, whereas fog forms when the air is in contact with the ground, which cools it and condenses the water vapor. The atmospheric phenomenon of fog decreases visibility to distances of less than 1 mile (1.6 km) and can affect ground, maritime, and air traffic. When the fog is light, it is called mist. In this case, visibility is reduced to 2 miles (3.2 km).

160 feet
(50 m)

The densest fog affects visibility at this distance and has repercussions on car, boat, and airplane traffic. In many cases, visibility can be zero.

OROGRAPHIC BARRIER
Fog develops on lee-side mountain slopes at high altitudes and occurs when the air becomes saturated with moisture.

4 OROGRAPHIC FOG

DEW
The condensation of water vapor on objects that have radiated enough heat to decrease their temperature below the dew point

Fog and Visibility

Visibility is defined as a measure of an observer's ability to recognize objects at a distance through the atmosphere. It is expressed in miles and indicates the visual limit imposed by the presence of fog, mist, dust, smoke, or any type of artificial or natural precipitation in the atmosphere. The different degrees of fog density have various effects on maritime, land, and air traffic.

DENSE FOG	THICK FOG	FOG	MIST

Means of transport are affected by visibility.

160 feet (50 m)	660 feet (200 m)	0.6 mile (1 km)	1.2 miles (2 km)

Types of Fog

Radiation fog forms during cold nights when the land loses the heat that was absorbed during the day. Frontal fog forms when water that is falling has a higher temperature than the surrounding air; the drops of rain evaporate, and the air tends to become saturated. These fogs are thick and persistent. Advection fog occurs when humid, warm air flows over a surface so cold that it causes the water vapor from the air to condense.

2 FRONTAL FOG
Formed ahead of a warm front.

1 RADIATION FOG
This fog appears only on the ground and is caused by radiation cooling of the Earth's surface.

FOG

FOG

3 ADVECTION FOG
Formed when a mass of humid and cool air moves over a surface that is colder than the air

FOG

The air becomes saturated as it ascends.

ASCENDING AIR

Warm air

BLOCKED FOG

High landmasses

Wind

MIST
Mist consists of salt and other dry particles imperceptible to the naked eye. When the concentration of these particles is very high, the clarity, color, texture, and form of objects we see are diminished.

INVERSION FOG
When a current of warm, humid air flows over the cold water of an ocean or lake, an inversion fog can form. The warm air is cooled by the water, and its moisture condenses into droplets. The warm air traps the cooled air below it, near the surface. High coastal landmasses prevent this type of fog from penetrating very far inland.

6 miles
(10 km)
Normal visibility.

1.9 miles
(3 km)

Brief Flash

Electrical storms are produced in large cumulonimbus-type clouds, which typically bring heavy rains in addition to lightning and thunder. The storms form in areas of low pressure, where the air is warm and less dense than the surrounding atmosphere. Inside the cloud, an enormous electrical charge accumulates, which is then discharged with a zigzag flash between the cloud and the ground, between the cloud and the air, or between one cloud and another. This is how the flash of lightning is unleashed. Moreover, the heat that is released during the discharge generates an expansion and contraction of the air that is called thunder.

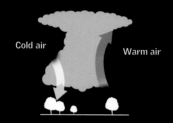

Cold air   Warm air

1 ORIGIN
Lightning originates within large cumulonimbus storm clouds. Lightning bolts can have negative or positive electric charges.

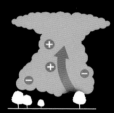

2 INSIDE THE CLOUD
Electrical charges are produced from the collisions between ice or hail crystals. Warm air currents rise, causing the charges in the cloud to shift.

SEPARATION
The charges become separated, with the positive charges accumulating at the top of the cloud and the negative charges at the base.

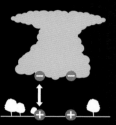

3 ELECTRICAL CHARGES
The cloud's negative charges are attracted to the positive charges of the ground. The difference in electrical potential between the two regions produces the discharge.

INDUCED CHARGE
The negative charge of the base of the cloud induces a positive charge in the ground below it.

TYPES OF LIGHTNING
Lightning can be distinguished primarily by the path taken by the electrical charges that cause them.

CLOUD-TO-AIR
The electricity moves from the cloud toward an air mass of opposite charge.

CLOUD-TO-CLOUD
A lightning flash can occur within a cloud or between two oppositely charged areas.

CLOUD-TO-GROUND
Negative charges of the cloud are attracted by the positive charges of the ground.

Lightning Rods

A lightning rod is an instrument whose purpose is to attract a lightning bolt and channel the electrical discharge to the ground so that it does no harm to buildings or people. A famous experiment by Benjamin Franklin led to the invention of this apparatus. During a lightning storm, he flew a kite into clouds, and it received a strong discharge. That marked the birth of the lightning rod, which consists of an iron rod placed on the highest point of the object to be protected and connected to the ground by a metallic, insulated conductor. The principle of all lightning rods, which terminate in one or more points, is to attract and conduct the lightning bolt to the ground.

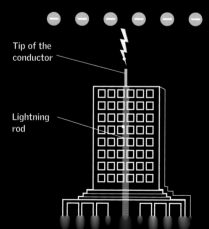

Tip of the conductor

Lightning rod

The primary function of lightning rods is to facilitate the electrostatic discharge, which follows the path of least electrical resistance.

THUNDER
This is the sound produced by the air when it expands very rapidly, generating shock waves as it is heated.

Cold air → Very hot air ← Very hot air ← Cold air

DISCHARGE
4
The discharge takes place from the cloud toward the ground after the stepped leader, a channel of ionized air, extends down to the ground.

8,700 miles per second (140,000 km/s) speed

Lightning bolt: 8,700 miles per second (140,000 km/s)

Airplane: 0.2 mile per second (0.3 km/s)

F1 car: 0.06 miles per second (0.1 km/s)

100 million volts
is the electrical potential of a lightning bolt.

A windmill generates 200 volts.

110 volts is consumed by a lamp.

RETURN STROKE
5
In the final phase, the discharge rises from the Earth to the cloud.

DISCHARGE SEQUENCE

channel	1st phase	2nd phase	3rd phase

1st return	2nd return	3rd return

A
The lightning bolt propagates through an ionized channel that branches out to reach the ground. Electrical charges run along the same channel in the opposite direction.

B
If the cloud has additional electrical charges, they are propagated to the ground through the channel of the first stroke and generate a second return stroke toward the cloud.

C
This discharge, as in the second stroke, does not have branches. When the return discharge ceases, the lightning flash sequence comes to an end.

POINT OF IMPACT

65 feet (20 m)
This is the radius of a lightning bolt's effective range on the surface of the Earth.

Lethal Force

Tornadoes are the most violent storms of nature. They are generated by electrical storms (or sometimes as the result of a hurricane), and they take the form of powerful funnel-shaped whirlwinds that extend from the sky to the ground. In these storms, moving air is mixed with soil and other matter rotating at velocities as high as 300 miles per hour (480 km/h). They can uproot trees, destroy buildings, and turn harmless objects into deadly airborne projectiles. A tornado can devastate a whole neighborhood within seconds.

Where and When

Most tornadoes occur in agricultural areas. The humidity and heat of the spring and summer are required to feed the storms that produce them. In order to grow, crops require both the humidity and temperature variations associated with the seasons.

Tornadoes
Agricultural areas

300 miles per hour (480 km/h)

Maximum velocity the tornado winds can attain

1,000

tornadoes are generated on average annually in the United States.

3:00 P.M.–9:00 P.M.

The period of the day with the highest probability of tornado formation

How They Form

Tornadoes begin to form when a current of warm air ascends inside a cumulonimbus cloud and begins to rotate under the influence of winds in the upper part of the cloud. From the base of the column, air is sucked toward the inside of the turning spiral. The air rotates faster as it approaches the center of the column. This increases the force of the ascending current, and the column continues to grow until it stretches from high in the clouds to the ground. Because of their short duration, they are difficult to study and predict.

FUJITA SCALE The Fujita-Pearson scale was created by Theodore Fujita to classify tornadoes according to the damage caused by the wind, from the lightest to the most severe.	**WIND VELOCITY MILES PER HOUR (KM/H)**	40-72 (64-116)	73-112 (117-180)	113-157 (181-253)	158-206 (254-332)	207-260 (333-418)	261-320 (420-512)
	CATEGORY	F0	F1	F2	F3	F4	F5
	EFFECTS	Damage to chimneys, tree branches broken	Mobile homes ripped from their foundations	Mobile homes destroyed, trees felled	Roofs and walls demolished, cars and trains overturned	Solidly built walls blown down	Houses uprooted from their foundations and dragged great distances

125 miles
(200 km)

The length of the path along the ground over which a tornado can move.

TOP
The top of the tornado remains inside the cloud.

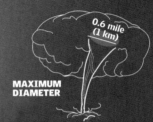

MAXIMUM DIAMETER

0.6 mile (1 km)

6 miles
(10 km)

Maximum height that it can attain.

1 BEGINNING OF A TORNADO
When the winds meet, they cause the air to rotate in a clockwise direction in the Southern Hemisphere and in the reverse direction in the Northern Hemisphere.

Strong wind

Mild Wind

Spinning funnel of air

VORTEX
Column of air that forms the lower part of a tornado; a funnel that generates violent winds and draws in air. It usually acquires the dark color of the dust it sucks up from the ground, but it can be invisible.

MULTIPLE VORTICES
Some tornadoes have a number of vortices.

2 ROTATION
The circulation of the air causes a decrease in pressure at the center of the storm, creating a central column of air.

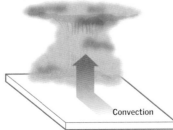

Convection

3 DESCENT
The central whirling column continues to descend within the cloud, perforating it in the direction of the ground.

Warm and humid wind

Cumulonimbus

Cold and dry wind

4 THE OUTCOME
The tornado reaches the Earth and depending on its intensity can send the roofs of buildings flying.

SPIRALING WINDS
First a cloud funnel appears that can then extend to touch the ground.

Storm

Humid wind

PATH
Normally the tornado path is no more than 160 to 330 feet (50-100 m) wide.

The tornado generally moves from the southwest to the northeast.

Some tornadoes are so powerful that they can rip the roofs off houses.

Anatomy of a Hurricane

A hurricane, with its ferocious winds, banks of clouds, and torrential rains, is the most spectacular meteorological phenomenon of the Earth's weather. It is characterized by an intense low-pressure center surrounded by cloud bands arranged in spiral form; these rotate around the eye of the hurricane in a clockwise direction in the Southern Hemisphere and in the opposite direction in the Northern Hemisphere. While tornadoes are brief and relatively limited, hurricanes are enormous and slow-moving, and their passage usually takes many lives.

DAY 1
A jumble of clouds is formed.

1 BIRTH
Forms over warm seas, aided by winds in opposing directions, high temperatures, humidity, and the rotation of the Earth.

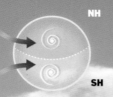

NH

Hurricanes in the Northern Hemisphere rotate counterclockwise, and those in the Southern Hemisphere rotate clockwise.

SH

FRINGES OF STORM CLOUDS
rotate violently around the central zone.

THE EYE
Central area, has very low pressure

Descending air currents

80° F
(27° C)

is likely the minimum sea-surface temperature capable of sustaining tropical cyclones.

The air wraps around the eye.

Strong ascendant currents

Cloud bands in the form of a spiral

EYE WALL
The strongest winds are formed.

VAPOR
Rises warm from the sea, forming a column of clouds. It rises 3,900 feet (1,200 m) in the center of the storm.

The trade winds are pulled toward the storm.

DAY 2
The clouds begin to rotate.

DAY 3
The spiral form becomes more defined.

TYPHOON

HURRICANE

Equator

CYCLONE

DANGER ZONE
The areas that are vulnerable to hurricanes in the United States include the Atlantic coast and the coast along the Gulf of Mexico, from Texas to Maine. The Caribbean and the tropical areas of the western Pacific, including Hawaii, Guam, American Samoa, and Saipan, are also zones frequented by hurricanes.

2 DEVELOPMENT
Begins to ascend, twisting in a spiral around a low-pressure zone.

DAY 6
Now mature, it displays a visible eye.

DAY 12
The hurricane begins to break apart when it makes landfall.

19 mph (30 km/h)
velocity at which it approaches the coast.

3 DEATH
As they pass from the sea to the land, they cause enormous damage. Hurricanes gradually dissipate over land from the lack of water vapor.

The high-altitude winds blow from outside the storm.

FRICTION
When the hurricane reaches the mainland, it moves more slowly; it is very destructive in this stage, since it is here that populated cities are located.

PATH OF THE HURRICANE

92 feet/high (28 m)
maximum height reached by the waves

1
2
3
4
5

CLASSIFICATION OF DAMAGE DONE
Saffir-Simpson category

	Damage	Speed mph (km/h)	High Tide feet (m)
CLASS 1	**Minimum**	74 to 95 (119 to 153)	4 to 5 (1.2 to 1.5)
CLASS 2	**Moderate**	96 to 110 (154 to 177)	6 to 8 (1.8 to 2.4)
CLASS 3	**Extensive**	111 to 130 (178 to 209)	9 to 12 (2.7 to 3.6)
CLASS 4	**Extreme**	131 to 155 (210 to 250)	13 to 18 (3.9 to 5.4)
CLASS 5	**Catastrophic**	more than 155 (250)	more than 18 (5.4)

WIND ACTIVITY

The winds flow outward.

Light winds give it direction and permit it to grow.

BUTTERFLY
It belongs to a group of species from phylum Arthropoda, class Insecta, order Lepidoptera.

CHAPTER 5

ANIMAL
KINGDOM

But for the charismatic presence of vertebrates (fish, amphibians, reptiles, birds, and mammals), animals comprise a universe that is dominated by a surprising 90 percent majority of invertebrates. Differentiated from other life forms by their unique features, animals are formed from many cells, get the energy they need by consuming other living beings, and respond to external stimuli.

What Is a Mammal?

Mammals share a series of characteristics that distinguish their class: a body covered by hair, the birth of live young, and the feeding of newborns on milk produced by the females' mammary glands. All breathe through lungs, and all possess a closed, double circulatory system and the most developed nervous systems in the animal kingdom. The ability to maintain a constant body temperature has allowed them to spread out and conquer every corner of the Earth, from the coldest climates to hot deserts and from the mountains to oceans.

A Body for Every Environment

Skin covered with hair and sweat glands helps create and maintain a constant body temperature. At the same time, with eyes placed on each side of the head (monocular vision, with the sole exception of the primates, which have binocular vision), they are afforded important angles of sight. Limbs are either of the foot or chiridium type, with slight variations depending on the part of the foot used for walking. In aquatic mammals, the limbs have evolved into fins; in bats, into wings. Hunters have powerful claws, and unguligrades (such as horses) have strong hooves that support the whole body when running.

BOTTLENOSE DOLPHIN
Tursiops truncatus.

5,487
The number of mammal species estimated to exist on Earth.

HAIR
Body hair is unique to mammals and absent in other classes of animals. Sirenians, with little hair, and cetaceans are exceptions; in both cases, the absence of hair is a result of the mammal's adaptation to an aquatic environment.

DENTITION
The majority of mammals change dentition in their passage to adulthood. Teeth are specialized for each function: molars for chewing, canines for tearing, and incisors for gnawing. In rodents such as chipmunks, the teeth are renewed by continuous growth.

CHIPMUNK
Family *Sciuridae.*

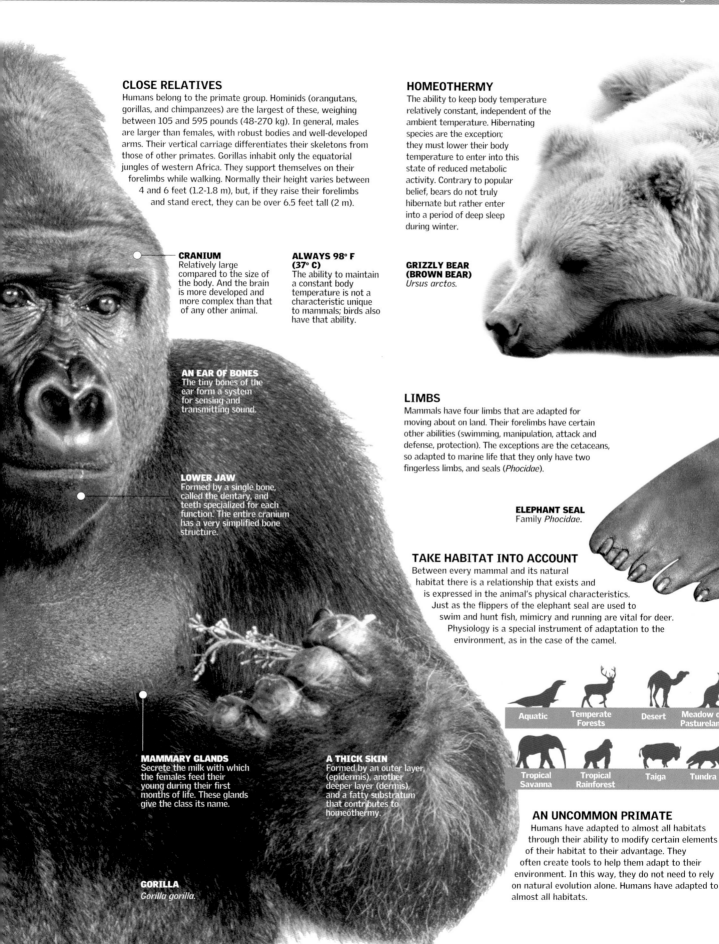

CLOSE RELATIVES

Humans belong to the primate group. Hominids (orangutans, gorillas, and chimpanzees) are the largest of these, weighing between 105 and 595 pounds (48-270 kg). In general, males are larger than females, with robust bodies and well-developed arms. Their vertical carriage differentiates their skeletons from those of other primates. Gorillas inhabit only the equatorial jungles of western Africa. They support themselves on their forelimbs while walking. Normally their height varies between 4 and 6 feet (1.2-1.8 m), but, if they raise their forelimbs and stand erect, they can be over 6.5 feet tall (2 m).

CRANIUM
Relatively large compared to the size of the body. And the brain is more developed and more complex than that of any other animal.

ALWAYS 98° F (37° C)
The ability to maintain a constant body temperature is not a characteristic unique to mammals; birds also have that ability.

AN EAR OF BONES
The tiny bones of the ear form a system for sensing and transmitting sound.

LOWER JAW
Formed by a single bone, called the dentary, and teeth specialized for each function. The entire cranium has a very simplified bone structure.

MAMMARY GLANDS
Secrete the milk with which the females feed their young during their first months of life. These glands give the class its name.

A THICK SKIN
Formed by an outer layer (epidermis), another deeper layer (dermis), and a fatty substratum that contributes to homeothermy.

GORILLA
Gorilla gorilla.

HOMEOTHERMY

The ability to keep body temperature relatively constant, independent of the ambient temperature. Hibernating species are the exception; they must lower their body temperature to enter into this state of reduced metabolic activity. Contrary to popular belief, bears do not truly hibernate but rather enter into a period of deep sleep during winter.

GRIZZLY BEAR (BROWN BEAR)
Ursus arctos.

LIMBS

Mammals have four limbs that are adapted for moving about on land. Their forelimbs have certain other abilities (swimming, manipulation, attack and defense, protection). The exceptions are the cetaceans, so adapted to marine life that they only have two fingerless limbs, and seals (*Phocidae*).

ELEPHANT SEAL
Family *Phocidae.*

TAKE HABITAT INTO ACCOUNT

Between every mammal and its natural habitat there is a relationship that exists and is expressed in the animal's physical characteristics. Just as the flippers of the elephant seal are used to swim and hunt fish, mimicry and running are vital for deer. Physiology is a special instrument of adaptation to the environment, as in the case of the camel.

Aquatic	Temperate Forests	Desert	Meadow or Pastureland
Tropical Savanna	Tropical Rainforest	Taiga	Tundra

AN UNCOMMON PRIMATE

Humans have adapted to almost all habitats through their ability to modify certain elements of their habitat to their advantage. They often create tools to help them adapt to their environment. In this way, they do not need to rely on natural evolution alone. Humans have adapted to almost all habitats.

Constant Heat

Mammals are homeothermic—which means they are capable of maintaining a stable internal body temperature despite environmental conditions. This ability has allowed them to establish themselves in every region of the planet. Homeostasis is achieved by a series of processes that tend to keep water levels and concentrations of minerals and glucose in the blood in equilibrium as well as prevent an accumulation of waste products—among other things.

Kings of the Arctic

The polar bear (or white bear) is a perfect example of adaptation to the inhospitable environment it inhabits. The fur, which can appear white, pale yellow, or cream in color, is actually translucent and colorless; it consists of two layers, one with thick short hairs and a superficial one with long hairs. The insulation necessary to survive in the Arctic is provided by its thick fur and by a layer of fat under the skin, both of which allow the polar bear to dive and swim in icy waters and to withstand blizzards.

GREAT SWIMMERS
Polar bears swim with ease in open waters and reach a speed of 6 miles an hour (10 km/h). They propel themselves with their great front paws and use their back feet as rudders. The bear's hair is hollow and filled with air, which helps with buoyancy. When the bear dives, its eyes remain open.

POLAR BEAR
Ursus maritimus.

SHELTERED CUBS

The cubs are born in winter, and the skin of the mother generates heat that protects the cubs from the extreme cold.

Migration

When spring begins, these bears travel south, escaping the breakup of the arctic ice.

METABOLISM

The layer of fat is between 4 and 6 inches (10-15 cm) thick and provides not only thermal insulation but also an energy reserve. When the temperature reaches critical levels—at the Pole it can drop to between -60° and -75° F (-50° to -60° C)—the animal's metabolism increases and begins to rapidly burn energy from fat and food. In this way, the polar bear maintains its body temperature.

UNDER THE ICE

Females dig a tunnel in the spring; when they become pregnant, they hibernate without eating and can lose 45 percent of their weight.

SECONDARY ACCESS TUNNEL

RESPIRATORY PATHWAYS

The bears have membranes in their snouts that warm and humidify the air before it reaches the lungs.

HAIR
An impermeable, translucent surface

Hollow chamber with air

CHAMBER OR REFUGE

MAIN ACCESS TUNNEL

ENTRANCE

LAYERS

GUARD HAIRS
Outer.

UNDERFUR
Inner.

FAT
4-6 inches (10-15 cm) thick.

CURLING UP

Many cold-climate mammals curl up into balls, covering their extremities and bending their tails over their bodies as a kind of blanket. In this way, the surface area subjected to heat loss will be minimal. Hot-climate animals stretch out their bodies to dissipate heat.

PRINCIPAL FAT RESERVES
Thighs, haunches, and abdomen.

over
6 mph (10 km/h)

is the average speed at which polar bears swim.

SLOW AND STEADY SWIMMING

Hind Legs
function as a rudder.

Forelimbs
function as a motor.

HYDRODYNAMIC ANATOMY

AND FINALLY...
THE FLOATING SLAB
When they tire of swimming, they rest, floating. They manage to cross distances of over 37 miles (60 km) in this manner.

TO GET OUT:
ANTISLIP PALMS
Their palms have surfaces with small papillae that create friction with ice, keeping them from slipping.

Grace and Movement

Horses, one of the odd-toed, hoofed, ungulate mammals, are considered symbols of grace and freedom. They have great vigor and can run swiftly because their spine bends very little, preventing unnecessary expenditure of energy during the rising and falling of their body mass. They are equipped with strong, light, and flexible bones, and their muscles work by contraction, arranged in pairs or groups that pull in opposing directions.

Power to Run

Horses are one of the most powerful mammals and achieve great speeds relative to their body mass. The natural purpose of their musculature is to allow them to flee their enemies. This ability has allowed the species to survive for millions of years. Their great energy is generated by contracting muscles.

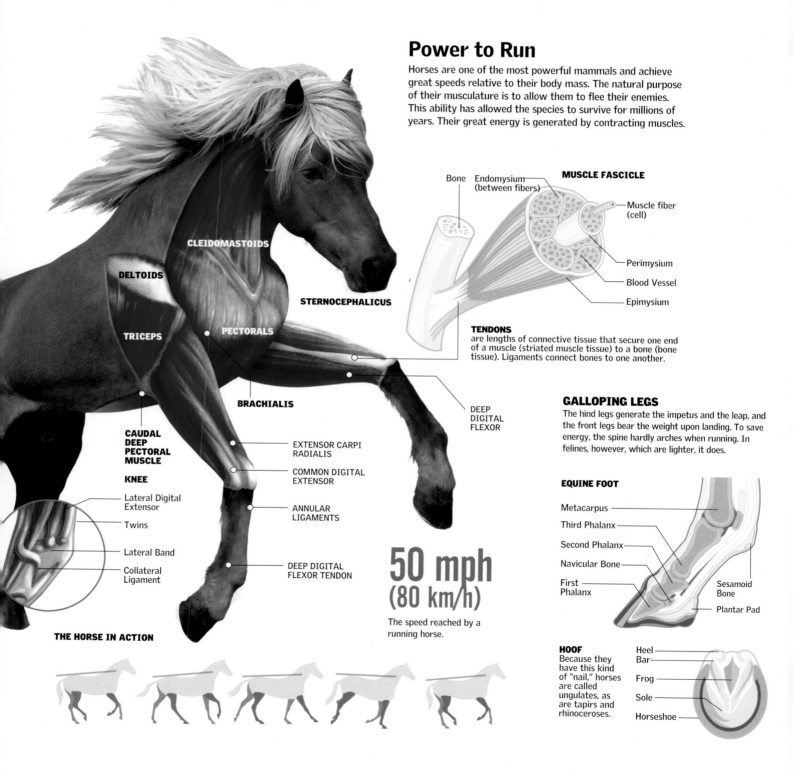

Bone   Endomysium (between fibers)

MUSCLE FASCICLE

Muscle fiber (cell)

Perimysium

Blood Vessel

Epimysium

CLEIDOMASTOIDS

DELTOIDS

STERNOCEPHALICUS

TRICEPS

PECTORALS

BRACHIALIS

TENDONS
are lengths of connective tissue that secure one end of a muscle (striated muscle tissue) to a bone (bone tissue). Ligaments connect bones to one another.

DEEP DIGITAL FLEXOR

CAUDAL DEEP PECTORAL MUSCLE

KNEE

Lateral Digital Extensor

Twins

Lateral Band

Collateral Ligament

EXTENSOR CARPI RADIALIS

COMMON DIGITAL EXTENSOR

ANNULAR LIGAMENTS

DEEP DIGITAL FLEXOR TENDON

50 mph (80 km/h)

The speed reached by a running horse.

THE HORSE IN ACTION

GALLOPING LEGS

The hind legs generate the impetus and the leap, and the front legs bear the weight upon landing. To save energy, the spine hardly arches when running. In felines, however, which are lighter, it does.

EQUINE FOOT

Metacarpus

Third Phalanx

Second Phalanx

Navicular Bone

First Phalanx

Sesamoid Bone

Plantar Pad

HOOF
Because they have this kind of "nail," horses are called ungulates, as are tapirs and rhinoceroses.

Heel

Bar

Frog

Sole

Horseshoe

Skeleton

34 bones in the cranium.

14 teeth in each maxillary bone, including:
3 molars
3 premolars
6 incisors
2 canines

- **ORBITAL CAVITY**
- **NASAL CAVITY**
- **BUCCAL CAVITY**

VERTEBRAE

7 CERVICAL

FROM 17 TO 19 DORSAL
Normally there are 18, but the number is often higher or lower.

Correct position of an equestrian

ATLAS
First cervical vertebra is articulated, allowing the nape to bend up and down.

Atlas

AXIS
Second cervical vertebra allows lateral movement—necessary for the horse to turn.

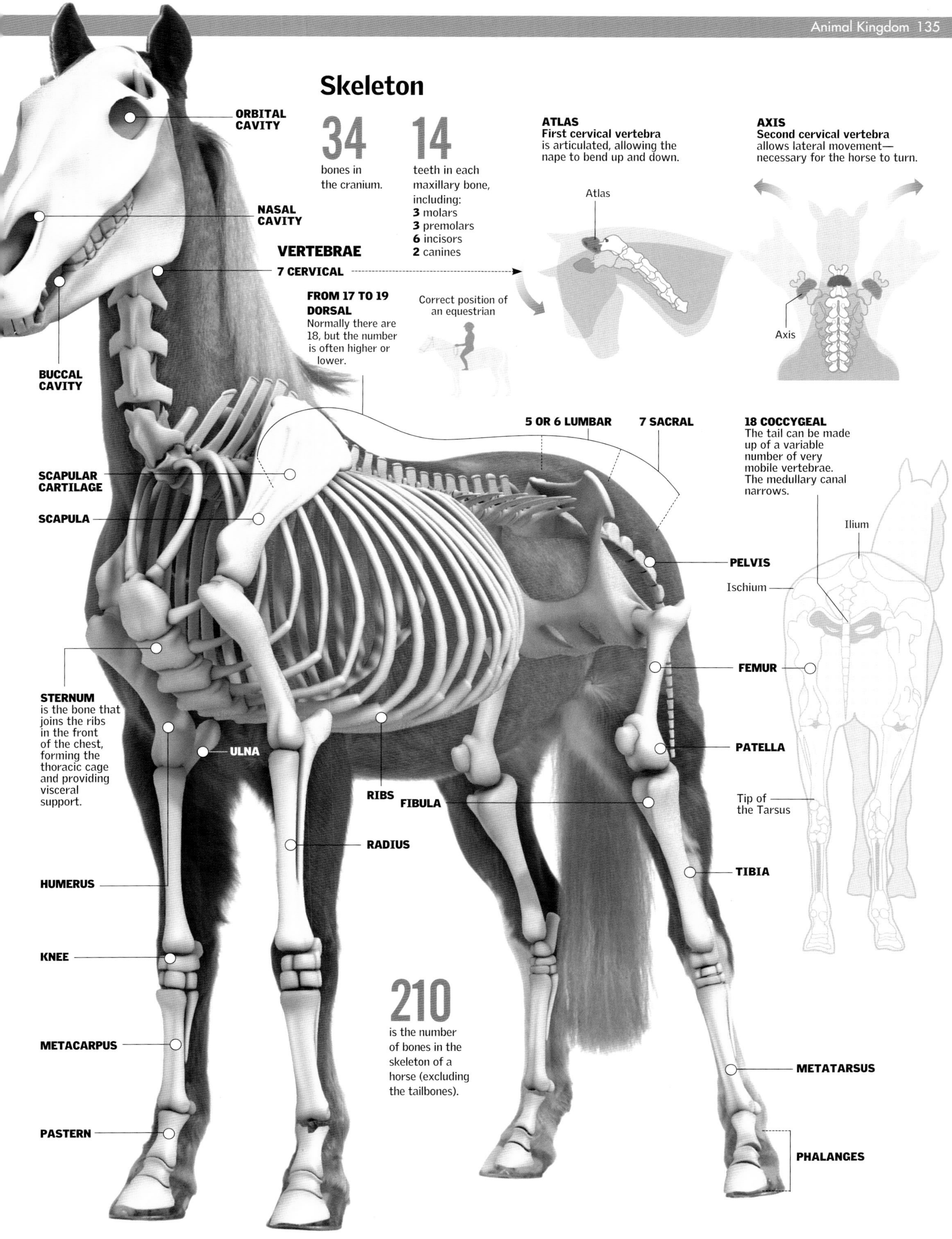

Axis

5 OR 6 LUMBAR   **7 SACRAL**

18 COCCYGEAL
The tail can be made up of a variable number of very mobile vertebrae. The medullary canal narrows.

Ilium

Ischium

- **SCAPULAR CARTILAGE**
- **SCAPULA**

- **PELVIS**

- **FEMUR**

- **STERNUM** is the bone that joins the ribs in the front of the chest, forming the thoracic cage and providing visceral support.

- **ULNA**

- **RIBS**
- **FIBULA**

- **RADIUS**

- **PATELLA**

Tip of the Tarsus

- **TIBIA**

- **HUMERUS**

- **KNEE**

210 is the number of bones in the skeleton of a horse (excluding the tailbones).

- **METACARPUS**

- **METATARSUS**

- **PASTERN**

- **PHALANGES**

Extremities

Mammals' extremities are basically either of the foot or chiridium type but modified according to the way in which each species moves about. Thus, for example, they become fins for swimming in aquatic mammals and membranous wings in bats. In land mammals, these variations depend on the way the animal bears its weight in walking: those that use the whole foot are called plantigrades; those that place their weight on their digits, digitigrades; and those that only touch the ground with the tips of their phalanges, ungulates.

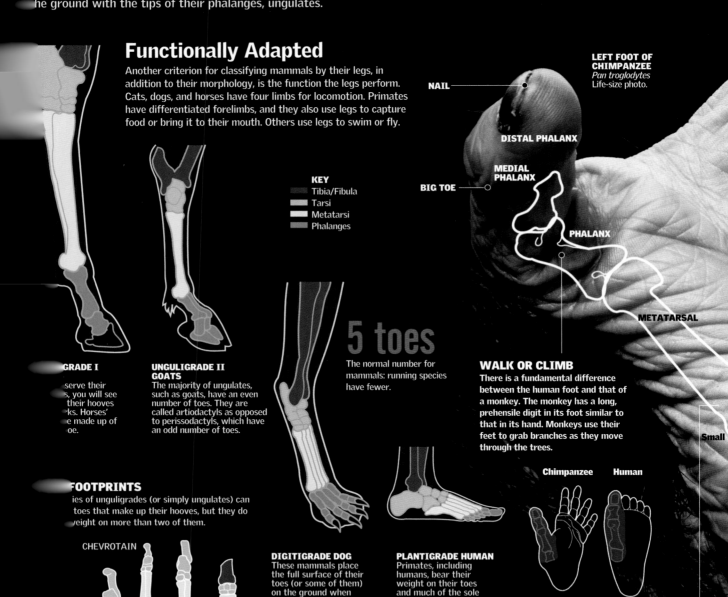

Functionally Adapted

Another criterion for classifying mammals by their legs, in addition to their morphology, is the function the legs perform. Cats, dogs, and horses have four limbs for locomotion. Primates have differentiated forelimbs, and they also use legs to capture food or bring it to their mouth. Others use legs to swim or fly.

KEY
- Tibia/Fibula
- Tarsi
- Metatarsi
- Phalanges

LEFT FOOT OF CHIMPANZEE
Pan troglodytes
Life-size photo.

SECOND TOE

NAIL

DISTAL PHALANX

MEDIAL PHALANX

BIG TOE

PHALANX

METATARSAL

Small

TARSI

5 toes
The normal number for mammals: running species have fewer.

WALK OR CLIMB
There is a fundamental difference between the human foot and that of a monkey. The monkey has a long, prehensile digit in its foot similar to that in its hand. Monkeys use their feet to grab branches as they move through the trees.

Chimpanzee   Human

GRADE I
serve their
s, you will see
their hooves
ks. Horses'
e made up of
oe.

UNGULIGRADE II GOATS
The majority of ungulates, such as goats, have an even number of toes. They are called artiodactyls as opposed to perissodactyls, which have an odd number of toes.

FOOTPRINTS
ies of unguligrades (or simply ungulates) can
toes that make up their hooves, but they do
weight on more than two of them.

CHEVROTAIN

DIGITIGRADE DOG
These mammals place the full surface of their toes (or some of them) on the ground when walking. They usually leave the mark of their front toes and a small part of the forefoot as a footprint. Dogs and cats are the best-known examples.

PLANTIGRADE HUMAN
Primates, including humans, bear their weight on their toes and much of the sole of the foot when walking, particularly on the metatarsus. Rats, weasels, bears, rabbits, skunks, raccoons, mice, and hedgehogs are also plantigrades.

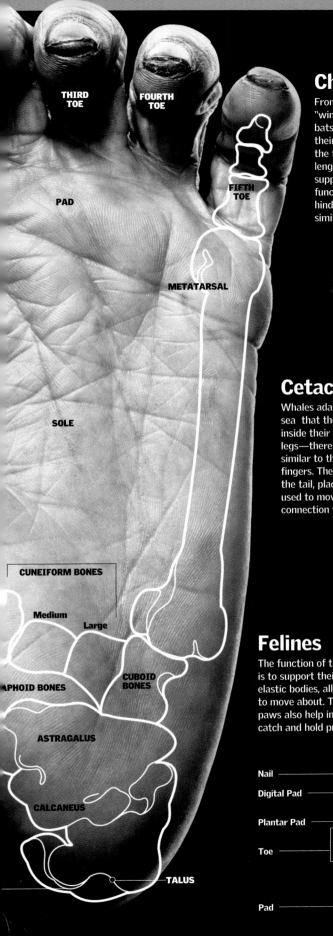

THIRD TOE

FOURTH TOE

FIFTH TOE

PAD

METATARSAL

SOLE

CUNEIFORM BONES

Medium

Large

PHOID BONES

CUBOID BONES

ASTRAGALUS

CALCANEUS

TALUS

Chiroptera

From the Greek, meaning "winged hand," this is how bats are designated because their forelimbs are modified, the fingers thinning and lengthening to be able to support a membrane that functions as a wing. The hind limbs did not change similarly: they have claws.

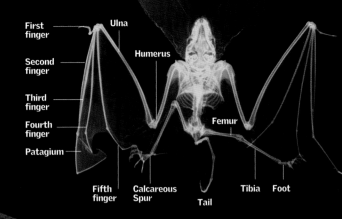

First finger

Ulna

Second finger

Humerus

Third finger

Fourth finger

Femur

Patagium

Fifth finger

Calcareous Spur

Tail

Tibia

Foot

Cetaceans

Whales adapted so well to the sea that they seem to be fish. But inside their fins —modified front legs—there is a bony structure similar to that of a hand with fingers. They have no hind limbs: the tail, placed horizontally and used to move in the water, has no connection to those limbs.

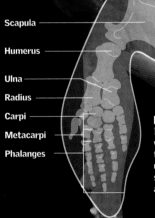

Tail

horizontal in mammals that swim, as distinct from fish.

Scapula

Humerus

Ulna

Radius

Carpi

Metacarpi

Phalanges

EVOLUTION

It is thought that whales descend from ancient marine ungulates, whose spines undulated up and down.

Felines

The function of their paws is to support their agile and elastic bodies, allowing them to move about. The front paws also help in hunting to catch and hold prey.

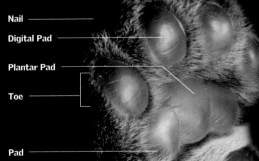

Nail

Digital Pad

Plantar Pad

Toe

Pad

RETRACTABLE NAIL

Phalanx

ELASTIC LIGAMENT
When the tendon contracts, this ligament retracts, and then the nail does, too.

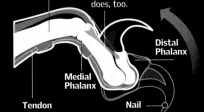

Distal Phalanx

Medial Phalanx

Tendon

Nail

Developed Senses

Dogs have inherited from wolves great hearing and an excellent sense of smell. Both perform an essential role in their relationship to their surroundings and many of their social activities. However, they are very dependent on the keenness of their senses depending on the habitat in which they develop. Whereas humans often remember other people as images, dogs do so with their sense of smell, their most important sense. They have 44 times more olfactory cells than people do, and they can perceive smells in an area covering some 24 square inches (150 sq cm). Dogs can discern one molecule out of a million other ones, and they can hear sounds so low that they are imperceptible to people.

Hearing

The auditory ability of dogs is four times greater than that of human beings, and it is highly developed. Their ability depends on the shape and orientation of their ears, which allow them to locate and pay closer attention to sounds, although this varies by breed. They can hear sharper tones and much softer sounds, and they can directly locate the spatial reference point where a noise was produced. Dogs hear sounds of up to 40 kilohertz, whereas the upper limit for human hearing is 18 kilohertz.

INSIDE THE COCHLEA

Reissner's Membrane

Scala Vestibuli

Organ of Corti

Scala Tympani

AURICULAR CARTILAGE

AUDITORY CANAL

MIDDLE EAR

COCHLEA

LABYRINTH

SEMICIRCULAR CANALS

AUDITORY NERVE

AUDITORY OSSICLES

Incus (anvil)

Malleus (hammer)

Stapes (stirrup)

COCHLEAR NERVE

AUDITORY CANAL

TYMPANIC MEMBRANE

Dome

Crest

Ciliary Cells

COCHLEA

INTERNAL STRUCTURE OF THE BULLA

The dome diverts sounds toward the bulla and other organs that direct electric signals to the brain.

OVAL WINDOW

EUSTACHIAN TUBE

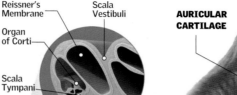

AUDITORY LEVELS

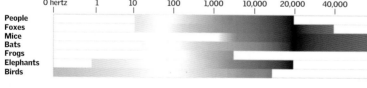

	0 hertz	1	10	100	1,000	10,000	20,000	40,000
People								
Foxes								
Mice								
Bats								
Frogs								
Elephants								
Birds								

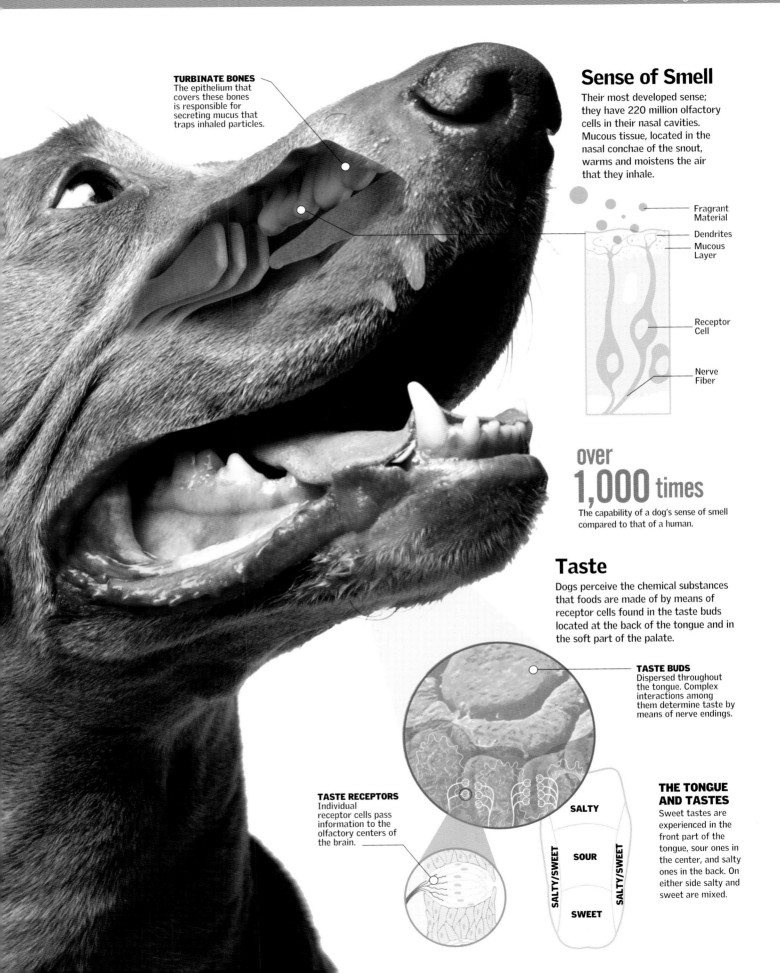

TURBINATE BONES
The epithelium that covers these bones is responsible for secreting mucus that traps inhaled particles.

Sense of Smell

Their most developed sense; they have 220 million olfactory cells in their nasal cavities. Mucous tissue, located in the nasal conchae of the snout, warms and moistens the air that they inhale.

Fragrant Material

Dendrites

Mucous Layer

Receptor Cell

Nerve Fiber

over 1,000 times

The capability of a dog's sense of smell compared to that of a human.

Taste

Dogs perceive the chemical substances that foods are made of by means of receptor cells found in the taste buds located at the back of the tongue and in the soft part of the palate.

TASTE BUDS
Dispersed throughout the tongue. Complex interactions among them determine taste by means of nerve endings.

TASTE RECEPTORS
Individual receptor cells pass information to the olfactory centers of the brain.

SALTY

SALTY/SWEET

SOUR

SALTY/SWEET

SWEET

THE TONGUE AND TASTES
Sweet tastes are experienced in the front part of the tongue, sour ones in the center, and salty ones in the back. On either side salty and sweet are mixed.

Looks That Kill

Tigers are the largest of the world's felines. Predators par excellence, they have physical skills and highly developed senses that they use to hunt for prey. Their daytime vision is as good as that of humans, except for a difficulty in seeing details. However, at night, when tigers usually hunt, their vision is six times keener than that of a human being, because tigers' eyes have larger anterior chambers and lenses and wider pupils.

Seeing Even in the Dark

Hunting animals depend on the keenness of their senses to detect their prey. Felines can dilate their pupils up to three times more than humans, and they see best when light is dim and their prey's movements are very subtle. A system of 15 layers of cells forms a sort of mirror (tapetum lucidum) located behind the retina or back of the eye. This mirror amplifies the light that enters and is also the reason that the animal's eyes shine in the dark. At the same time, their eyes are six times more sensitive to light than those of people. Tigers' nocturnal vision also increases because of the great adaptability of their circular pupils when they are completely open.

BINOCULAR VISION
Part of the field of vision of one eye overlaps that of the other eye, which makes three-dimensional vision possible. Hunters' skills depend on binocular vision, because it allows them to judge the distance and size of their prey.

FOCUS 1

FOCUS 2

Tigers have a 255° angle of vision, of which 120° is binocular, whereas humans have 210° with 120° of it binocular.

50 times
The light amplification capability of the retina of felines

FIELD OF VISION

Right Field of Vision

Left Field

Binocular Field

PUPILS
They regulate the passage of light to the retina by contracting in bright light and dilating in the dark. In each species of mammal, the pupils have a distinctive shape.

TIGER   CAT   GOAT

RETINA

LENS

IRIS

CORNEA

PUPIL

CONJUNCTIVA

VITREOUS
HUMOR

OPTIC
NERVE

ROD   CONE

**RETINA OF
A DIURNAL
ANIMAL**
Cones, which
distinguish colors
and details,
along with light,
predominate.

**RETINA OF A
NOCTURNAL
ANIMAL**
Rods, super-
sensitive
to light,
predominate.

LIGHTS OR COLORS

The retina's
sensitivity to light
depends on rod-
shaped cells, and
forms and colors
depend on other
cells, which are cone-
shaped. In tigers, the
former predominate.

Field of Vision

HUMAN

DOG WITH LONG
SNOUT

SHORT-SNOUTED DOG

HARE

Herbivores

Ruminants, such as cows, sheep, or deer, have stomachs made of four chambers with which they carry out a unique kind of digestion. Because these animals need to eat large quantities of grass in very short times—or else be easy targets for predators!—they have developed a digestive system that allows them to swallow food, store it, and then return it to the mouth to chew calmly. When animals carry out this activity, they are said to ruminate.

KEY

- ▬ INGESTION AND FERMENTATION
- ▬ RUMINATION
- ▬ REABSORPTION OF NUTRIENTS
- ▬ ACID DIGESTION
- ▬ DIGESTION AND ABSORPTION
- ▬ FERMENTATION AND DIGESTION

Teeth

Herbivorous animals such as horses and bovids have molars with a large flat surface that reduces food to pulp, as well as incisors for cutting grass. Grinding is also done by the molars.

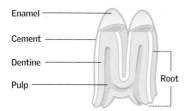

- Enamel
- Cement
- Dentine
- Pulp
- Root

Cows wrap their tongues around the food.

Then they chew it with lateral movements.

RETICULUM

1 Cows lightly chew grass and ingest it into their first two stomachs: the rumen and the reticulum. Food passes continually from the rumen to the reticulum (nearly once every minute). There various bacteria colonies begin fermenting the food.

2 When cows feel satiated, they regurgitate balls of food from the rumen and chew them again in the mouth. This is called rumination; it stimulates salivation, and, as digestion is a very slow process, cows make use of rumination to improve their own digestion together with the intervention of anaerobic microorganisms such as protozoa, bacteria, and fungi.

40 gallons (150 liters)

of saliva are produced daily in the process.

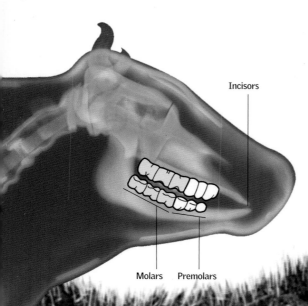

- Incisors
- Molars
- Premolars

THE RUMINATION PROCESS

Helps ruminants reduce the size of the ingested food particles. It is part of the process that allows them to obtain energy from plant cell walls, also called fiber.

 A REGURGITATION   B REMASTICATION   C REINSALIVATION   D REINGESTION

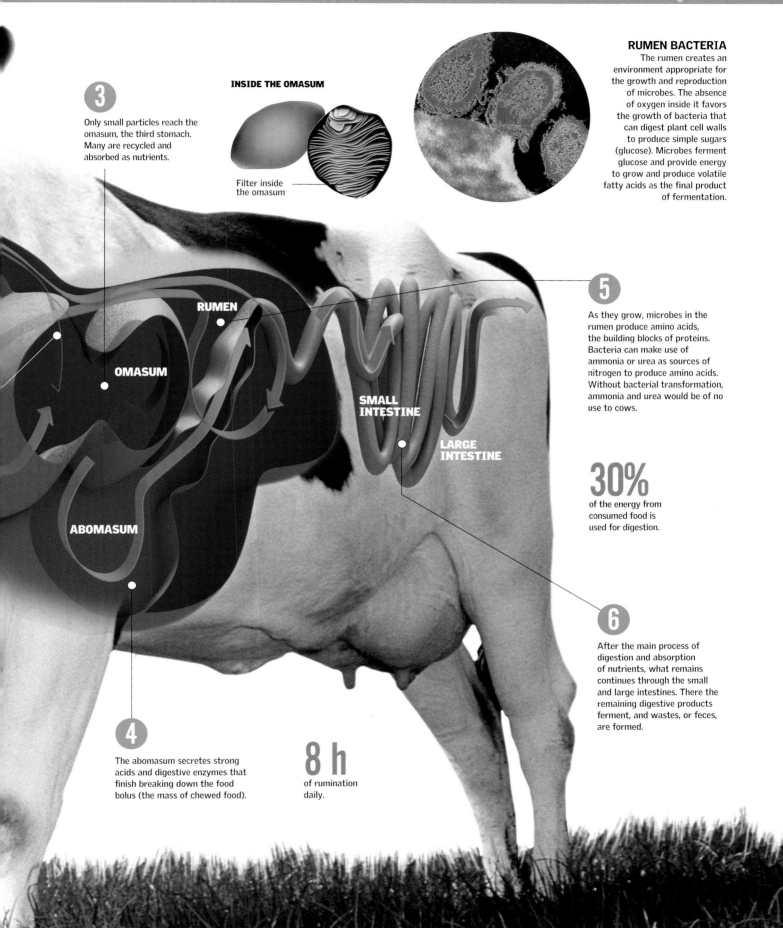

3

Only small particles reach the omasum, the third stomach. Many are recycled and absorbed as nutrients.

INSIDE THE OMASUM

Filter inside the omasum

RUMEN BACTERIA

The rumen creates an environment appropriate for the growth and reproduction of microbes. The absence of oxygen inside it favors the growth of bacteria that can digest plant cell walls to produce simple sugars (glucose). Microbes ferment glucose and provide energy to grow and produce volatile fatty acids as the final product of fermentation.

RUMEN

OMASUM

SMALL INTESTINE

LARGE INTESTINE

ABOMASUM

5

As they grow, microbes in the rumen produce amino acids, the building blocks of proteins. Bacteria can make use of ammonia or urea as sources of nitrogen to produce amino acids. Without bacterial transformation, ammonia and urea would be of no use to cows.

30%

of the energy from consumed food is used for digestion.

6

After the main process of digestion and absorption of nutrients, what remains continues through the small and large intestines. There the remaining digestive products ferment, and wastes, or feces, are formed.

4

The abomasum secretes strong acids and digestive enzymes that finish breaking down the food bolus (the mass of chewed food).

8 h

of rumination daily.

Deep Sleep

How many times have you heard the expression "dead as a dormouse"? The comparison is no accident, although it should be understood that dormice do not die: they merely hibernate. In the cold season, low temperatures and scarcity of food lead many mammals to enter into lethargic states. Body temperatures drop, heart rates and respiration slow down, and they lose consciousness.

HAZEL DORMOUSE
Muscardinus avellanarius.

Habitat	Almost all Europe
Habits	Hibernate 4 months of the year
Gestation	22 to 28 days

Weight
2 ounces
(51 g)

4 to 7 inches
(10-17 cm)

Their tails are very long. They can measure up to 5 inches (13.5 cm) long.

When Active

The energy they consume during hibernation is obtained from the subcutaneous fat layer built up during the autumn. Their nutrition comes from leaves, bark, nuts, and other (mainly plant) foods. Before the arrival of winter, they stock up on dried fruits to increase their energy, allowing them to easily climb trees and walls. Before hibernating, they spend all their time eating, accumulating reserves for winter.

95°F (35°C)
Their normal body temperature.

2 BALL
Dormice begin to form a ball out of these materials, in imitation of the posture they will adopt during hibernation.

1 RAW MATERIALS
To build their nests, dormice collect twigs, leaves, moss, feathers, and hair.

11 ounces (300 g)
is what they can weigh after accumulating fat reserves before hibernating.

LEAVES OF THE OAK TREES
Dormice are very fond of oak trees.

8 months
They are conscious and active.

March

NUTS
Although they consume snails and insects, dormice begin to feed on nuts prior to hibernation.

CHESTNUT
Its caloric contribution increases their energy reserves.

ACORNS
The nuts of oak trees (genus *Quercus*) are a favorite food of dormice.

Building the Nest

Dormice build their nests out of twigs, moss, and leaves, although they can also hibernate in trees, stone walls, or old buildings, creating a nest from fur, feathers, and leaves. They then settle into the nest, forming a ball. When they cannot find a natural refuge, dormice may settle into birds' nests with total impunity.

③
HOLLOW BALL
Like an ovenbird nest, the ball must be hollow so it can shelter the dormouse.

④
FINISHED NEST
With an entrance in front, the hollow ball has been transformed into a nest.

November

December

February

4 months
They remain in a state of hibernation.

50%
Weight loss after consuming all their reserves.

Hibernation

During this period, dormice enter into a deep sleep. Body temperature drops to 34° F (1° C), appreciably decreasing the heart rate. In fact, up to 50 minutes can transpire between breaths. Throughout these months, they slowly use up their reserves, losing up to 50 percent of their body weight. Their endocrine system is almost totally at rest: the thyroid ceases functioning, as does the interstitial tissue of the testicles.

34°F (1°C)
Their body temperature during hibernation.

POSITION OF THE BODY

Tail
They cover part of the body with it.

Head
They hide it behind their long tail.

Feet
remain flexed during these months.

Respiration
Fifty minutes can pass between breaths.

Energy
They obtain it from the subcutaneous fat reserves they accumulated in the fall.

Heart
Heartbeats decrease considerably.

BIORHYTHM OF A DORMOUSE WHILE HIBERNATING

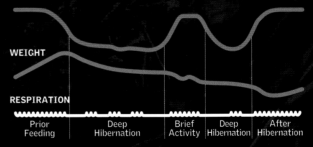

TEMPERATURE

WEIGHT

RESPIRATION

| Prior Feeding | Deep Hibernation | Brief Activity | Deep Hibernation | After Hibernation |

OTHER PLACES FOR HIBERNATION

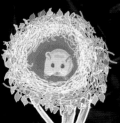

BIRD'S NEST
If they do not find a place to build their nest, they may take over a bird's nest.

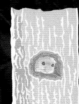

HOLE IN A TREE
can also serve as a burrow for hibernation.

Record Breath-Holders

Sperm whales are unique animals whose species is remarkable for many reasons. On the one hand, they have the ability to dive to a maximum depth of 9,800 feet (3,000 m) and remain underwater without oxygen for up to two hours. They are able to do this by means of a complex physiological mechanism that, for example, can decrease their heart rate, store and use air in the muscles, and prioritize the delivery of oxygen to certain vital organs such as the heart and lungs. They are the largest whales with teeth, which are found only on the lower mandible.

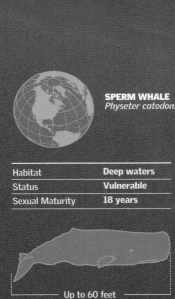

SPERM WHALE
Physeter catodon.

Habitat	Deep waters
Status	Vulnerable
Sexual Maturity	18 years

Up to 60 feet
(18 m)

Weight

20 to 90 tons

By Comparison

11 elephants of 8 tons apiece

BLOWHOLE

Up to 120 minutes
is the length of time they can spend underwater without breathing.

1 SPIRACLE
The sperm whale breathes oxygen into its body through spiracles located on the top of its head.

2 REPRIORITIZING OXYGEN
Sperm whales can allocate oxygen to certain vital organs, such as the lungs and heart, directing it away from the digestive system.

MOUTH
Because of the placement of the nostrils, sperm whales can swim with their mouth open and capture prey. They feed on squid.

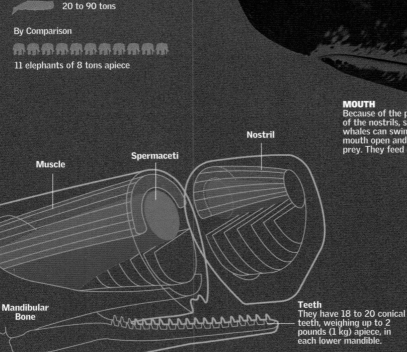

Muscle

Spermaceti

Nostril

Mandibular Bone

Teeth
They have 18 to 20 conical teeth, weighing up to 2 pounds (1 kg) apiece, in each lower mandible.

Spermaceti Organ

Sperm whales' ability to dive to great depths could be due in part to their spermaceti organ, located in their heads. It consists of a large mass of waxy oil that helps them both float and take deep dives. Its density changes with temperature and pressure change. It, like the melon of a dolphin, directs sound, focusing clicks, since its eyes are of little use when far from light.

COMPOSITION

90% Spermaceti Oil
It is made up of esters and triglycerides.

Adaptation in Respiration

When they dive to great depths, sperm whales activate an entire physiological mechanism that makes maximum use of their oxygen reserves. This produces what is called a thoracic and pulmonary collapse, causing air to pass from the lungs to the trachea, reducing the absorption of the toxin nitrogen. They also rapidly transmit nitrogen from the blood to the lungs at the end of the dive, thus reducing the circulation of blood to the muscles. Sperm whales' muscles contain a large amount of myoglobin, a protein that stores oxygen, allowing the whales to stay underwater much longer.

Blowhole
Upon submerging, it fills with water, which cools the spermaceti oil and makes it denser.

Heart
The heart rate slows down during the dive, limiting oxygen consumption.

Blood
An ample blood flow, rich in hemoglobin, transports elevated levels of oxygen to the body and brain.

ON THE SURFACE
Blowhole remains open, allowing the whales to breathe as much oxygen as they can before diving.

WHEN THEY DIVE
powerful muscles tightly close the opening of the blowhole, keeping water from entering.

Retia Mirabilia
The retia is a network of blood vessels (mirabilia) that filter the blood entering the brain.

Lungs
absorb oxygen very efficiently.

TAIL
is large and horizontal and is the whale's main means of propulsion.

3 BRADYCARDIA
During a dive, the heart rate drops (a condition known as bradycardia), which lowers oxygen consumption.

Dive

True diving champions, sperm whales can dive to depths of 9,800 feet (3,000 m), descending up to 10 feet (3 m) per second in search of squid. As a general rule, their dives last about 50 minutes, but they can remain underwater up to two hours. Before beginning a deep dive, they lift their caudal fin completely out of the water. They do not have a dorsal fin, but they do have a few triangular humps on the posterior part of their body.

0 FEET (0 M)
ON THE SURFACE
They inhale oxygen through the blowhole located at the top of the head.

+ 3,300 FEET (1,000 M)
90 MINUTES
They store 90 percent of their oxygen in their muscles, so they can be submerged for a long time.

0 FEET (0 M)
ON THE SURFACE
They exhale all the air from their lungs; this is called spouting, or blowing.

Making Use of Oxygen

Sperm whales can dive deeper and stay submerged longer than any other mammal, because they have various ways of saving oxygen: an ability to store it in their muscles, a metabolism that can function anaerobically, and the inducement of bradycardia during a dive.

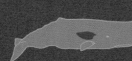

15%
amount of air replaced in one breath

85%
amount of air replaced in one breath

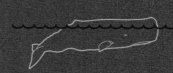

Nocturnal Flight

Bats are the only mammals that can fly. Scientists call them Chiroptera, a term derived from Greek words meaning "winged hands." Their forelimbs have been transformed into hands with very long fingers joined together by a membrane (called the patagium) that forms the surface of the wing. These mammals' senses are so sensitive that they can move and hunt quickly and accurately in the dark.

Expert Pilots

Moved by their chest and back muscles, bats' wings push downward and backward, generating both thrust and lift. Then the wings spread sideways and upward. Finally they move forward until the tips almost rub the bat's head. Many of these flying mammals can drift through the air, gliding without flapping and maneuvering by folding their wings.

...heir Radar

...st of the time bats fly at night ...ear-total darkness. Instead of ..., they use a natural system ...lar to sonar or radar to guide ...nselves. This system makes ...of acoustical signals the bats ...nselves emit while flying. This ...em allows them to recognize ...ocation of any object in front ...em or of prey, along with its ...ction, size, or speed. It is as if ...were seeing without light

1 The animal emits an acoustical vibration imperceptible to the human ear because of its high frequency (about 18 kHz). The signal strikes the objects around it.

2 When the signals bounce back, the bat perceives their intensity and phase difference—the faster and more intense the return signal, the nearer the

Hibernation

These bats spend the winter in a lethargic state hanging by their feet, faces down, in caves and other dark places. Bats are warm-blooded animals while they are active and become similar to cold-blooded creatures when they are asleep. They enter into a state of hibernation more rapidly and easily than any other mammal, and they can survive in cold temperatures for many months—even inside refrigerators—without needing to feed.

FRUIT BAT (FRANQUET'S EPAULETTED BAT)
Epomops franqueti.

Habitat	Forests of Ghana and Congo
Family	Pteropodae
Length of wingspan	14 inches (36 cm)

60 miles per hour (97 km)
The speed some bats may reach during flight.

HUMERUS

RADIUS

THUMB

SECOND FINGER

FOURTH FINGER

THIRD FINGER

PATAGIUM

Flexible Wings

The patagium is formed by the membranes between the digits. In some species, the wings are also extended by an additional membrane (uropatagium), which joins the hind limbs to the tail. Their wings are not only used for flying (pushing the air as if they were oars in water) but also help to maintain a constant body temperature and to trap insects, upon which bats feed.

Hand or Wing
The first finger, or thumb, has no membrane and is used as a claw. Powerful muscles move the entire wing.

UROPATAGIUM

Elastic Fibers
The texture of the wing is soft and flexible. It is lined with blood vessels.

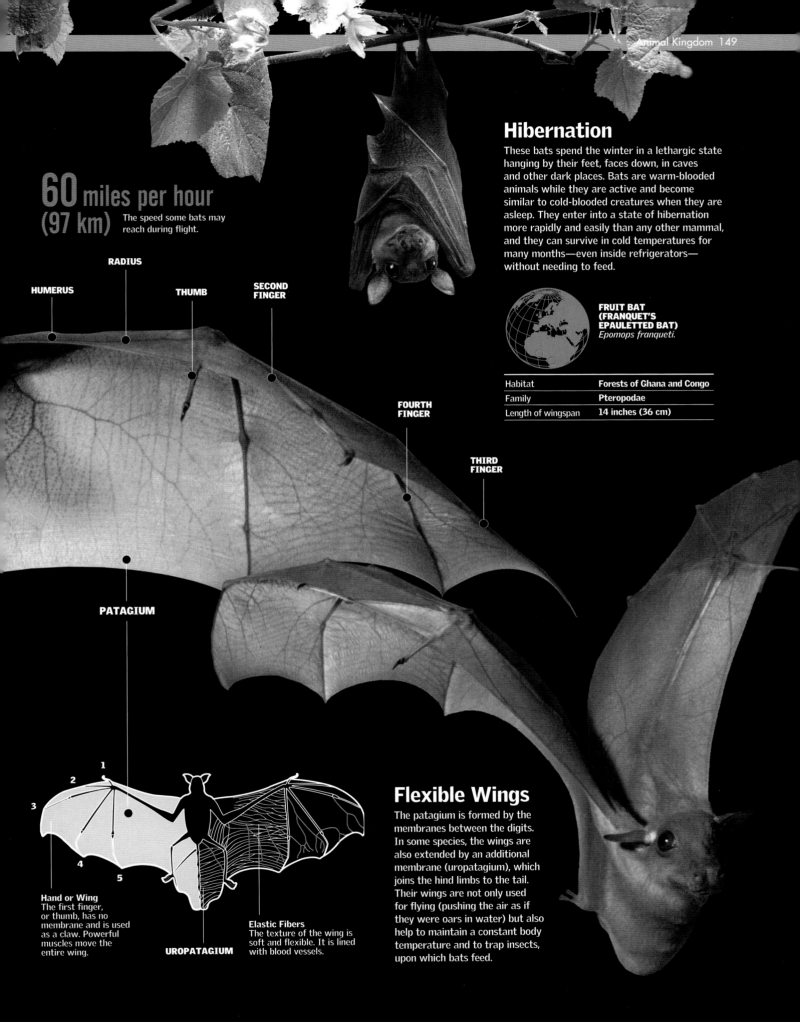

The Language of Water

The ways in which cetaceans communicate with others of their kind are among the most sophisticated in the animal kingdom. Dolphins, for example, click with their mandibles when in trouble and whistle repeatedly when afraid or excited. During courtship and mating, they touch and caress. They also communicate through visual signals—such as leaping—to show that food is close by. They have a wide variety of ways to transmit important information.

HAVING FUN
Play for dolphins, as with other mammals, fulfills an essential role in the formation of social strata.

BOTTLENOSE DOLPHIN
Tursiops truncatus.

Family	Delphinidae
Adult Weight	330 to 1,400 pounds (150 to 650 kg)
Longevity	30 to 40 years

7 to 13 feet (2-4 m)

They reach
22 mph (35 km/h)

MELON
is an organ filled with low-density lipids that concentrate and direct the pulses emitted, sending waves forward. The shape of the melon can be varied to better focus the sounds.

SPIRACLE   **LIP**

NASAL AIR SAC

DORSAL FIN
allows dolphins to maintain their equilibrium in the water.

CAUDAL FIN
has a horizontal axis (unlike that of fish), which serves to propel dolphins forward.

PECTORAL FIN

LARYNX

1 EMISSION
Sounds are generated by air passing through the respiratory chambers. But it is in the melon that resonance is generated and amplified. Greater frequencies and intensities are achieved in this way.

HOW THE SOUND IS PRODUCED

1 INHALATION
The spiracle opens so oxygen can enter.

Spiracle

Air to the lungs

2 The nasal air sacs begin to inflate.

They can go 12 minutes without taking in oxygen.

Air in the lungs

4 The nasal air sacs deflate

Melon

3 EXHALATION
Air resonates in the nasal sacs and the produced sound is directed through the melon.

Sound

Brain

MANDIBLE
The lower mandible plays a very important role in the transmission of sounds to the inner ear.

3 RECEPTION AND INTERPRETATION
The middle ear sends the message to the brain. Dolphins hear frequencies from 100 Hz up to 150 kHz (the human ear can hear only up to 15 kHz). Low-frequency signals (whistles, snores, grunts, clicking) are key in the social life of dolphins, cetaceans that cannot live alone.

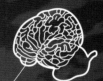

3 pounds (1.4 kg)
Human brain.

4 pounds (1.7 kg)
Dolphin brain.

MORE NEURONS
A dolphin's brain, which processes the signals, has at least double the convolutions of those of humans, as well as nearly 50 percent more neurons.

MIDDLE EAR

2 MESSAGE
Low-frequency signals are used for communication with other dolphins, and high-frequency signals are used as sonar.

1 mile per second (1.5 km/s)
Sound waves travel 4.5 times faster in water than in air.

ECHOLOCATION

A The dolphin emits a series of **clicking** sounds from the nasal cavity.

B The melon concentrates the clicks and projects them forward.

C These waves bounce off objects they encounter in their way.

E The intensity, pitch, and return time of the echo indicate the size, position, and direction of the obstacle.

D Part of the signal bounces back and returns to the dolphin in the form of an echo.

SIGNAL WITH ECHO

Click   Click

Echo   Echo

0 s   6 s   12 s   18 s

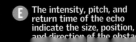

Beyond Feathers

Defining what a bird is brings to mind an animal covered with feathers that has a toothless bill and anterior extremities morphed into wings. Other distinguishing characteristics are that they are warm-blooded and have pneumatic bones—bones filled with air chambers instead of marrow. Birds have very efficient circulatory and respiratory systems and great neuromuscular and sensory coordination.

Variety and Uniformity

We can find birds in every type of environment: aquatic, aerial, and terrestrial, in polar regions and in tropical zones. Their adaptation to the environment has been very successful. Nevertheless, birds are one of the groups that display the fewest differences among their members.

**0.06 ounces
(1.6 g)**
Weight of the smallest bird.

WINE-THROATED HUMMINGBIRD

AFRICAN OSTRICH

**330 pounds
(150 kg)**
Weight of the largest bird.

WHITE-THROATED SPARROW
A small bird that lives in North America and on the Iberian Peninsula

Adaptation to Flying

Some crucial anatomic and physiological characteristics explain birds' ability to fly. Their bodies and feathers reduce friction with the air and improve lift. Their strong muscles, light bones, air sacs, and closed double circulatory system also play a role in their ability to fly.

FEATHERS
Unique. No other living animal has them. They are appealing for their structure, variety, and constant renewal.

WINGS
propel, maintain, and guide birds during flight. They have modified bones and characteristic plumage.

**105.8° F
(41° C)**
is their body temperature.

PENGUIN

**-75 F
(-60° C)**
The temperature penguins can endure in Antarctica.

THIGH

FLIGHT FEATHERS

UNDERTAIL COVERTS

TARSUS

TAIL
The last vertebrae merge into the pygostyle. The tail feathers develop in this area.

STRUCTURE
Balance in movement. A bird's internal architecture contributes to its stability. The location of its feet and wings helps to concentrate its weight close to its center of gravity.

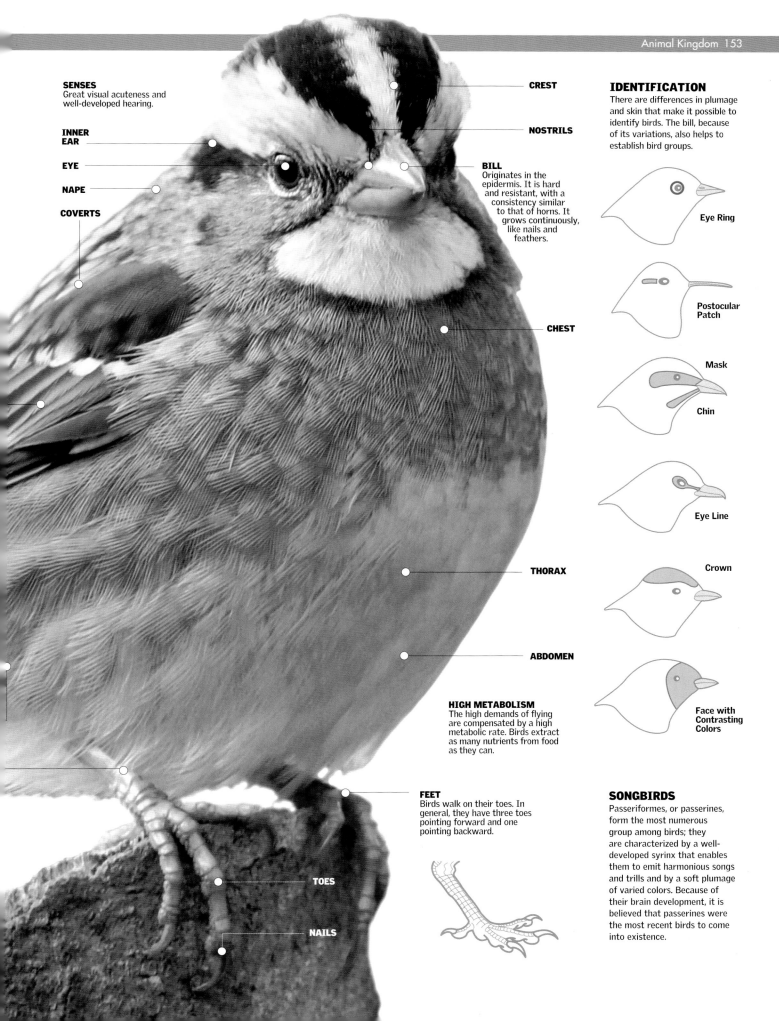

SENSES
Great visual acuteness and well-developed hearing.

INNER EAR

EYE

NAPE

COVERTS

CREST

NOSTRILS

BILL
Originates in the epidermis. It is hard and resistant, with a consistency similar to that of horns. It grows continuously, like nails and feathers.

CHEST

THORAX

ABDOMEN

HIGH METABOLISM
The high demands of flying are compensated by a high metabolic rate. Birds extract as many nutrients from food as they can.

FEET
Birds walk on their toes. In general, they have three toes pointing forward and one pointing backward.

TOES

NAILS

IDENTIFICATION
There are differences in plumage and skin that make it possible to identify birds. The bill, because of its variations, also helps to establish bird groups.

Eye Ring

Postocular Patch

Mask

Chin

Eye Line

Crown

Face with Contrasting Colors

SONGBIRDS
Passeriformes, or passerines, form the most numerous group among birds; they are characterized by a well-developed syrinx that enables them to emit harmonious songs and trills and by a soft plumage of varied colors. Because of their brain development, it is believed that passerines were the most recent birds to come into existence.

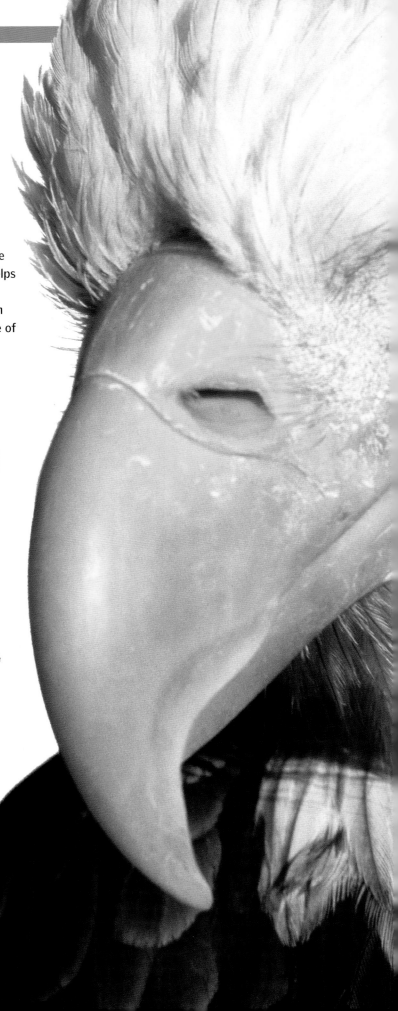

The Senses

In birds, the sense organs are concentrated on the head, except for the sense of touch, which is found all over the body. Birds have the largest eyes with respect to the size of their bodies. This enables them to see distant objects with considerable precision. Their field of vision is very broad, over 300 degrees, but in general they have little binocular vision. The ear—a simple orifice, but very refined in nocturnal hunters—helps them notice sounds inaudible to humans, which facilitates the detection of prey while flying. The senses of touch and smell, on the other hand, are important only to some birds, and the sense of taste is almost nonexistent.

The Ear

Birds' ears are simpler than those of mammals: a bird's ear has no outer portion, and in some cases it is covered with rigid feathers. A notable part of the ear is the columella—a bone that birds share with reptiles. The ear is nonetheless well developed, and birds have very acute hearing; whereas human beings can detect just one note, birds can detect many. The ear is essential to a bird's balance, a key factor in flying. It is also believed that in certain species the ear works as a barometer, indicating altitude.

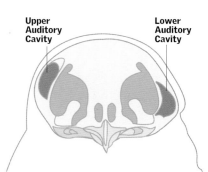

Upper Auditory Cavity **Lower Auditory Cavity**

LOCATION OF THE EARS
Located at different heights on the head, the ears cause the sense of hearing to occur with a slight delay. In nocturnal hunters, such as owls, this asymmetry allows for the triangulation of sounds and the tracking of prey with a minimal margin of error.

Touch, Taste, and Smell

The sense of touch is well developed in the bill and tongue of many birds, especially in those birds that use them to find food, such as shore birds and woodpeckers. Usually the tongue is narrow, with few taste buds, but they are sufficient to distinguish among salty, sweet, bitter, and acidic tastes. The sense of smell is not very developed: although the cavity is broad, the olfactory epithelium is reduced. In some birds, such as kiwis and scavengers (condors, for example), the olfactory epithelium is more developed.

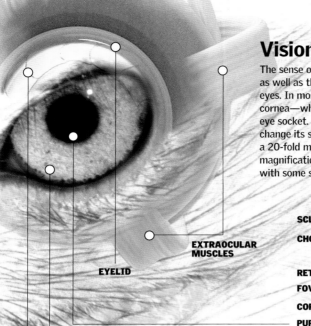

Vision

The sense of vision is the most developed sense in birds because some flight maneuvers, as well as the recognition of food from afar, depend on it. Birds have relatively large eyes. In most cases, they are wider than they are deep because the lens and the cornea—which is supported by a series of sclerotic bony plates—project beyond the eye socket. In hunting birds, the eyes are almost tubular. The muscles around the eye change its shape, alter the lens, and create greater visual acuity: birds typically have a 20-fold magnification (and sometimes, as in the case of some diving birds, a 60-fold magnification), in comparison with humans. Their sensitivity to light is also remarkable, with some species being able to recognize light spectra invisible to the human eye.

EXTRAOCULAR MUSCLES

EYELID

EXTRAOCULAR MUSCLES

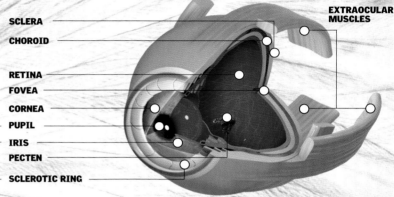

SCLERA
CHOROID
RETINA
FOVEA
CORNEA
PUPIL
IRIS
PECTEN
SCLEROTIC RING

FIELD OF VISION

The eyes—when located on the sides of the head, as is the case with most birds—create a broad field of vision: more than 300 degrees. Each eye covers different areas, focusing on the same object only when looking ahead through a narrow binocular field of vision.

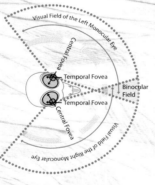

Visual Field of the Left Monocular Eye

Central Fovea

Temporal Fovea

Binocular Field

Temporal Fovea

Central Fovea

Visual Field of the Right Monocular Eye

THE HUMAN FIELD OF VISION

The eyes, located at the front, move together, covering the same area. Because human beings cannot move their eyes independently from each other, they have only binocular vision.

Visual Field of the Left Monocular Eye

Binocular Field

Visual Field of the Right Monocular Eye

COMPARISON OF BINOCULAR FIELDS

Binocular vision is essential for measuring distances without making mistakes. The brain processes the images that each eye generates separately as if they were a single image. The small differences between the two images allow the brain to create a third one in depth, or in three dimensions. Hunting birds, for which the correct perception of distance is a life-and-death matter, tend to have eyes located toward the front, with a wide field of binocular vision. In contrast, birds with lateral eyes calculate distance by moving their heads, but they record a larger total field of vision to avoid becoming prey. Owls are the birds with the greatest binocular vision—up to 70 degrees.

A   B

Monocular Field of Vision

HUNTING BIRDS' FIELD OF VISION

Frontal eyes reduce the total field of vision but allow for a wide field of binocular vision.

A   B

Monocular Field of Vision

Binocular Field of Vision

NONHUNTING BIRDS' FIELD OF VISION

The lateral eyes open the field of vision to as much as 360 degrees but reduce the binocular field.

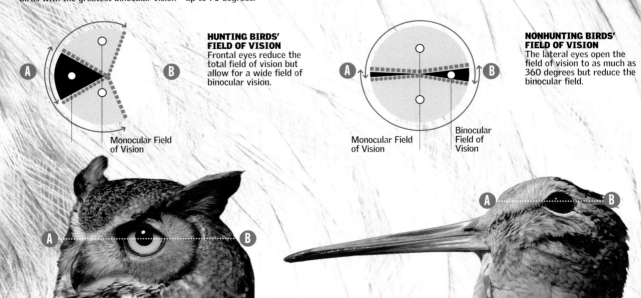

A .... B

A .... B

Wings to Fly

Wings are highly modified arms that, through their unique structure and shape, enable most birds to fly. There are many types of wings; they vary by species. For instance, penguins, which are flightless, use their wings for the specialized task of swimming. Among all wings that have existed in the animal kingdom, those of birds are the best for flying. Their wings are light and durable, and in some cases their shape and effectiveness can be modified during flight. To understand the relationship between wings and a bird's weight, the concept of wing loading, which helps explain the type of flight for each species, is useful.

Wings in the Animal Kingdom

Wings have always been modified arms, from the first models on pterosaurs to those on modern birds. Wings have evolved, beginning with the adaptation of bones. Non-avian wings have a membranous surface composed of flexible skin. They extend from the bones of the hand and body usually down to the legs, depending on the species. Avian wings, on the other hand, are based on a very different principle: the arm and hand form a complex of skin, bone, and muscle, with a wing surface consisting of feathers. Furthermore, the avian wing allows for important changes in form, depending on the bird's adaptation to the environment.

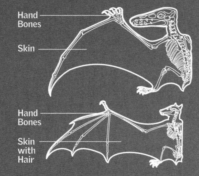

Hand Bones

Skin

PTERODACTYLS
Still had talons, and only one finger extended their wings.

Hand Bones

Skin with Hair

BATS
Four fingers extend the membrane, and the thumb remained as a talon.

Hand Bones

Feathers

BIRDS
The fused fingers form the tip of the wing where the rectrices, or primary feathers, are attached.

Types of Wings

According to the environment in which they live and the type of flight they perform, birds have different wing shapes that allow them to save energy and to perform efficiently during flight. The wing shape also depends on the bird's size. Consequently, the number of primary and secondary feathers changes depending on the needs of a given species.

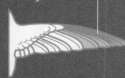

The external primary feathers are longer.

The outermost primary feathers are shorter than the central ones.

They are wide at the base, with separate feather tips.

There are many secondary feathers.

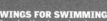

Short feathers are located all over the wing.

FAST WING
Remiges are large and tight to allow for flapping; the surface is reduced to prevent excessive friction.

ELLIPTICAL WINGS
Functional for mixed flights, they are very maneuverable. Many birds have them.

WINGS FOR SOARING ABOVE LAND
Wide, they are used to fly at low speeds. The separate remiges prevent turbulence when gliding.

WINGS FOR SOARING ABOVE THE OCEAN
Their great length and small width make them ideal for gliding against the wind, as flying requires.

WINGS FOR SWIMMING
In adapting to swimming, the feathers of penguins became short, and they serve primarily as insulation.

Wing Size and Loading

11.5 ft (3.5 m)

WANDERING ALBATROSS

5 ft (1.5 m)

The wingspan is the distance between the tips of the wings. Together with width, it determines the surface area, which is an essential measurement for bird flight. Not just any wing can support any bird. There is a close relationship between the animal's size (measured by weight) and the surface area of its wings. This relationship is called wing loading, and it is crucial in understanding the flight of certain species. Albatrosses, with large wings, have low wing loading, which makes them great gliders, whereas hummingbirds have to flap their small wings intensely to support their own weight.
The smaller the wing loading, the more a bird can glide; the bigger, the faster a bird can fly.

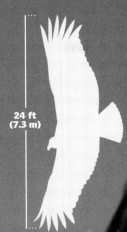

24 ft (7.3 m)

Argentavis magnificens (extinct)

LARGER FINGER

SMALLER FINGER

CARPOMETACARPUS

ALULAR DIGIT
Controls the alula, a feathered projection on the front edge of the wing.

ULNA

RADIUS

HUMERUS

CORACOID

STERNUM OR KEEL

PRIMARY COVERTS
They cover the remiges and, with the alula, change the wing shape at will.

PRIMARIES
They are in charge of propulsion; they are also called remiges.

MEDIAN WING COVERTS
They change the wing's lift when they rise slightly.

SECONDARIES
Their number varies greatly depending on the species. They complete the surface.

GREATER WING COVERTS
They create more surface area and cover the intersection point of the tertiaries.

TERTIARIES
Together with the secondaries, they create the wing's surface.

LOOSE FEATHERS
Sometimes barbicels are missing, and feathers on the wing come apart, creating a loose and ruffled appearance.

PRIMARY FEATHERS
Flying birds have from nine to 12 primary feathers. Running birds may have up to 16.

FUNCTION
The wings of ostriches carry out the functions of balancing, temperature regulation, and courtship.

Flightless Wings

Among these, penguins' wings are an extreme case of adaptation: designed for rowing underwater, they work as fins. On running birds, wings' first and foremost function is to provide balance as the bird runs. These wings are also related to courtship, as birds show off their ornamental feathers during mating season by opening their wings or flapping them. Wings are also very efficient at controlling temperature, as birds use them as fans to ventilate their bodies.

Fish anatomy

Most fish have the same internal organs as amphibians, reptiles, birds, and mammals. The skeleton acts as a support, and the brain receives information through the eyes and the lateral line to coordinate the motions of the muscles in propelling the fish through the water. Fish breathe with gills, they have a digestive system designed to transform food into nutrients, and they have a heart that pumps blood through a network of blood vessels.

SIMPLE EYE
Each eye focuses to one side; there is no binocular vision.

Suspensory ligament — Retina
Lens — Optic nerve
Iris —

BRAIN
receives information and coordinates all the fish's actions and functions.

Cyclostomata

Its digestive tract is little more than a straight tube extending from its round, jawless mouth to the anus. Because of their simplicity, many species of lampreys are parasites. They live off the blood of other fish and have thin pharyngeal sacs instead of gills.

45
The current number of species of cyclostomata

MOUTH

GILLS
Structures with multiple folds that provide oxygen to the blood

HEART
receives all the blood and pumps it toward the gills.

LIVER

CAUDAL FIN

ANUS

EYE

BREATHING SACS

HEART

LIVER

FIRST DORSAL FIN

LAMPREY
Lampetra sp.

INTESTINE

TOOTHED MOUTH

SUPPORT FOR PHARYNGEAL SACS

NOTOCHORD

TESTICLES

VERTEBRAE

RIGHT KIDNEY

STOMACH

BRAIN

GONAD

Chondrichthyes

A shark has the same organic structures as a bony fish, except for the swim bladder. A shark also has a corkscrew-like structure called a spiral valve at the end of its intestine to increase the surface area for absorption of nutrients.

NASAL PIT

SHARK
Carcharodon sp.

MOUTH

GILL SLITS

HEART

LIVER

STOMACH

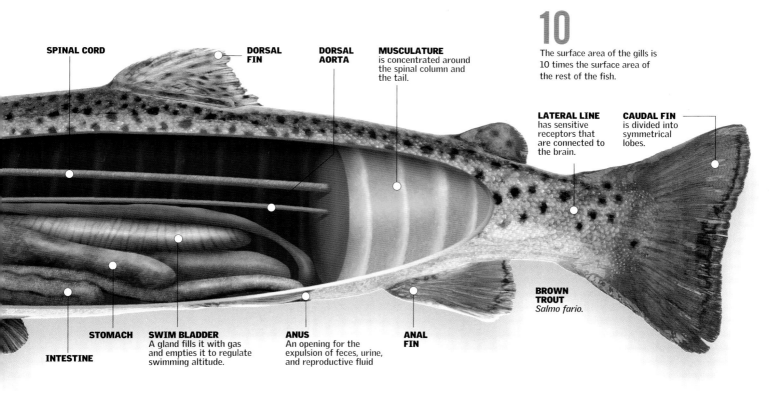

SPINAL CORD

DORSAL FIN

DORSAL AORTA

MUSCULATURE
is concentrated around the spinal column and the tail.

10
The surface area of the gills is 10 times the surface area of the rest of the fish.

LATERAL LINE
has sensitive receptors that are connected to the brain.

CAUDAL FIN
is divided into symmetrical lobes.

BROWN TROUT
Salmo fario.

STOMACH

INTESTINE

SWIM BLADDER
A gland fills it with gas and empties it to regulate swimming altitude.

ANUS
An opening for the expulsion of feces, urine, and reproductive fluid

ANAL FIN

Osteichthyes

Typically, their organs are compressed in the lower front quarter of the body. The rest of their internal structure consists mainly of the muscles that the fish uses to swim. Some bony fish, such as carp, have no stomach but rather a tightly coiled intestine.

REGULATION OF SALINITY

FRESHWATER FISH
Freshwater fish run the risk of losing salt to their environment. They drink only a small quantity of water, and they obtain additional salt from their food.

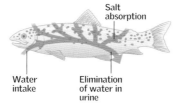

Salt absorption

Water intake

Elimination of water in urine

SALTWATER FISH
These fish constantly absorb salt water to replenish the water in their bodies, but they must eliminate excess salt from the marine environment.

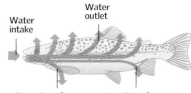

Water intake

Water outlet

Excretion of salts through the gills

Excretion of salts through urine

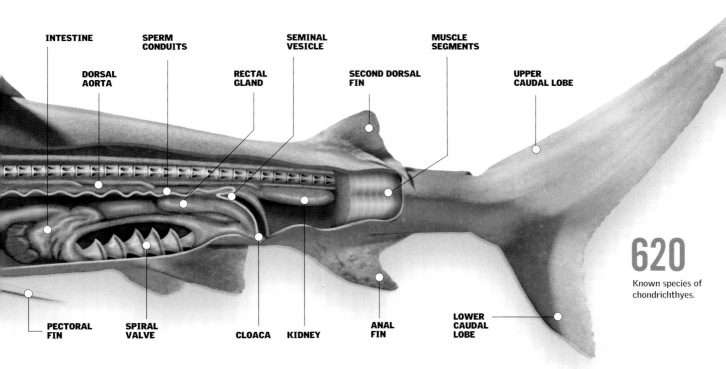

INTESTINE

DORSAL AORTA

SPERM CONDUITS

RECTAL GLAND

SEMINAL VESICLE

SECOND DORSAL FIN

MUSCLE SEGMENTS

UPPER CAUDAL LOBE

PECTORAL FIN

SPIRAL VALVE

CLOACA

KIDNEY

ANAL FIN

LOWER CAUDAL LOBE

620
Known species of chondrichthyes.

Protective Layer

Most fish are covered with scales, an external layer of transparent plates. All fish of a given species have the same number of scales. Depending on the family and genus of a fish, its scales can have a variety of characteristics. Scales on the lateral line of the body have small orifices that link the surface with a series of sensory cells and nerve endings. It is also possible to determine a fish's age by studying its scales.

FOSSILIZED SCALES
The remains of these thick, shiny, enameled scales belong to the extinct genus *Lepidotes*, a fish that lived during the Mesozoic Era.

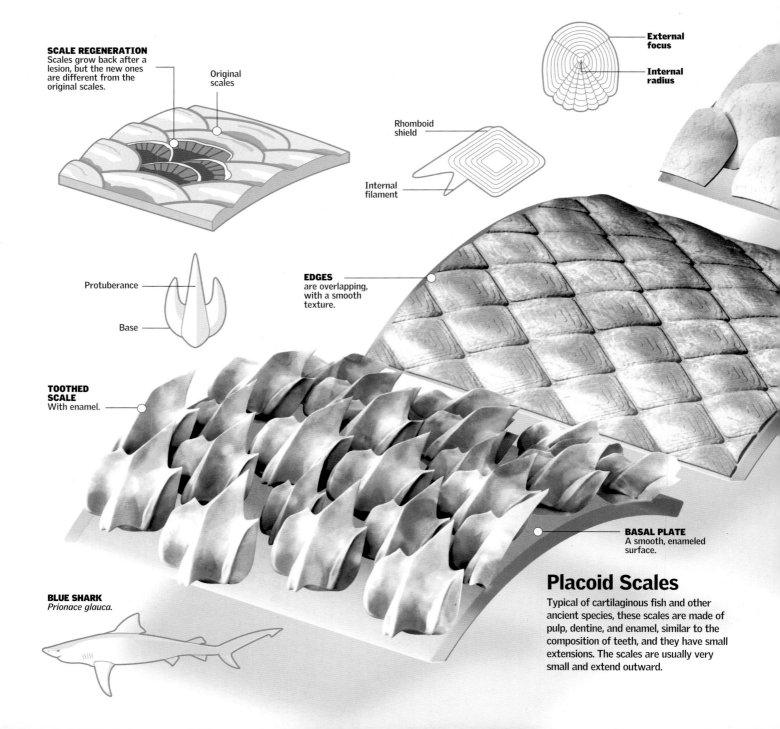

SCALE REGENERATION
Scales grow back after a lesion, but the new ones are different from the original scales.

Original scales

External focus

Internal radius

Rhomboid shield

Internal filament

Protuberance

Base

EDGES
are overlapping, with a smooth texture.

TOOTHED SCALE
With enamel.

BASAL PLATE
A smooth, enameled surface.

Placoid Scales

Typical of cartilaginous fish and other ancient species, these scales are made of pulp, dentine, and enamel, similar to the composition of teeth, and they have small extensions. The scales are usually very small and extend outward.

BLUE SHARK
Prionace glauca.

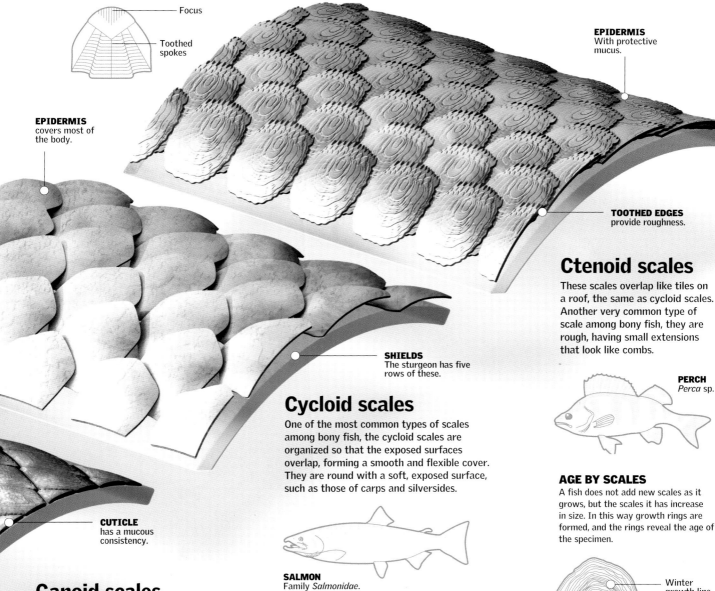

Focus

Toothed spokes

EPIDERMIS
covers most of
the body.

EPIDERMIS
With protective
mucus.

TOOTHED EDGES
provide roughness.

Ctenoid scales

These scales overlap like tiles on
a roof, the same as cycloid scales.
Another very common type of
scale among bony fish, they are
rough, having small extensions
that look like combs.

PERCH
Perca sp.

SHIELDS
The sturgeon has five
rows of these.

Cycloid scales

One of the most common types of scales
among bony fish, the cycloid scales are
organized so that the exposed surfaces
overlap, forming a smooth and flexible cover.
They are round with a soft, exposed surface,
such as those of carps and silversides.

AGE BY SCALES

A fish does not add new scales as it
grows, but the scales it has increase
in size. In this way growth rings are
formed, and the rings reveal the age of
the specimen.

CUTICLE
has a mucous
consistency.

SALMON
Family *Salmonidae*.

Winter
growth line

Summer
growth line

Exposed
area

Ganoid scales

Rhomboid in shape, these scales are
interwoven and connected with fibers. The
name comes from their outer covering, which
is a layer of ganoin, a type of shiny enamel.
Sturgeon and pipefish have scales of this type.

DISTRIBUTION OF SCALES

Most scales occur in rows that slant diagonally
downward and back. Species can be accurately
identified by the number of rows (as counted along
the lateral line), among other characteristics.

Transverse
line

Lateral
line

STURGEON
Acipenser sturio.

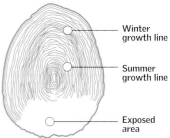

RED SNAPPER
Lutjanus campechanus.

The Art of Swimming

To swim, fish move in three dimensions: forward and back, left and right, and up and down. The main control surfaces that fish use for maneuvering are the fins, including the tail, or caudal fin. To change direction, the fish tilts the control surfaces at an angle to the water current. The fish must also keep its balance in the water; it accomplishes this by moving its paired and unpaired fins.

This fish swims upside down, seeking food sources that are less accessible to other species.

UPSIDE-DOWN CATFISH
Synodontis nigriventris.

MUSCLES
The tail has powerful muscles that enable it to move like an oar.

GREAT WHITE SHARK
Carcharodon carcharias.

Red muscles are for slow or regular movements.

Larger white muscles are for moving with speed, but they tire easily.

1

STARTING OUT
The movement of a fish through the water is like that of a slithering snake. Its body goes through a series of wavelike movements similar to an S curve. This process begins when the fish moves its head slightly from side to side.

The crest of the body's wave moves from back to front.

In its side-to-side movement, the tail displaces the water.

STREAMLINED SHAPE
Like the keel of a ship, the rounded contours of a fish are instrumental. In addition, most of a fish's volume is in the front part of its body. As the fish swims forward, its shape causes the density of the water ahead to be reduced relative to the density of the water behind. This reduces the water's resistance.

At first the tail is even with the head.

The head moves from side to side.

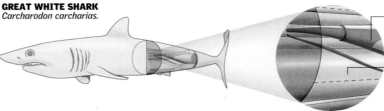

THE FISH'S KEEL
A ship has a heavy keel in the lower part to keep it from capsizing. Fish, on the other hand, have the keel on top. If the paired fins stop functioning to keep the fish balanced, the fish turns over because its heaviest part tends to sink, which happens when fish die.

Keel   Live fish   Dead fish

THE FASTEST

The powerful caudal fin displaces large amounts of water.

SAILFISH
Istiophorus platypterus.

The unfurled dorsal fin can be up to 150 percent of the width of the fish's body.

Its long upper jaw enables it to slice through the water, aiding this fish's hydrodynamics.

70 mph (109 km/h)
The maximum swimming speed it attains

FORWARD MOTION

Results from the synchronized S-curve movement of the muscles surrounding the spinal column. These muscles usually make alternating lateral motions. Fish with large pectoral fins use them like oars for propulsion.

The oarlike movement of the tail is the main force used for forward motion.

The dorsal fin keeps the fish upright.

The pectoral fins maintain balance and can act as brakes.

The ventral fins stabilize the fish for proper balance.

BALANCE

When the fish is moving slowly or is still in the water, the fins can be seen making small movements to keep the body in balance.

UPWARD AND DOWNWARD

The angle of the fins relative to the body allows the fish to move up or down. The paired fins, located in front of the center of gravity, are used for this upward or downward movement.

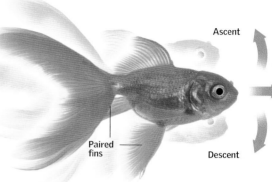

Ascent

Paired fins

Descent

2 FORCEFUL STROKE

Muscles on both sides of the spinal column, especially the tail muscles, contract in an alternating pattern. These contractions power the wavelike movement that propels the fish forward. The crest of the wave reaches the pelvic and dorsal fins.

The crest of the wave passes to the first dorsal fins.

When the crest reaches the area between the two dorsal fins, the tail fin begins its push to the right.

3 COMPLETE CYCLE

When the tail moves back toward the other side and reaches the far right, the head will once again turn to the right to begin a new cycle.

1 second

The amount of time it takes for this shark to complete one swimming cycle.

CAT SHARK
Scyliorhinus sp.

The resulting impulse moves the fish forward.

Swimming in Groups

Only bony fish can swim in highly coordinated groups. Schools of fish include thousands of individuals that move harmoniously as if they were a single fish. To coordinate their motion they use their sight, hearing, and lateral line senses. Swimming in groups has its advantages: it is harder to be caught by a predator, and it is easier to find companions or food.

School

A group of fish, usually of the same species, that swim together in a coordinated manner and with specific individual roles.

1 cubic mile (4 cu km)

The area that can be taken up by a school of herring.

The fish on the outside, guided by those in the middle, are in charge of keeping the group safe.

The fish in the middle control the school.

Between Land and Water

As indicated by their name (amphi, "both," and bios, "life"), these animals lead a double life. When young, they live in the water, and when they become adults they live outside it. In any case, many must remain near water or in very humid places to keep from drying out. This is because amphibians also breathe through their skin, and only moist skin can absorb oxygen. Some typical characteristics of adult frogs and toads include a tailless body, long hind limbs, and large eyes that often bulge.

Amphibian Anatomy

Amphibian anatomy has several peculiarities. Larvae, such as tadpoles, have a respiratory system with gills. Most species develop lungs when they reach adulthood. They also have a trachea, pharynx, and saclike lungs, even though skin breathing is at times more important than lung breathing. The heart has two auricles and one ventricle, and the digestive and excretory systems are similar to those of mammals.

VOCAL SACS
Both toads and frogs sing. Even though the sound is produced by their vocal cords, in males the sound is amplified by means of inflatable sacs on each side of the larynx.

THE SKIN
Amphibians breathe through their skin, which is clean and smooth, without hair or scales. They must always keep it moist, because it has a strong tendency to dry out. Even though they have mucous glands that help maintain moisture, amphibians must live in damp places. The skin of most amphibians protects them from possible predators and has poisonous glands that secrete unpleasant and even toxic substances.

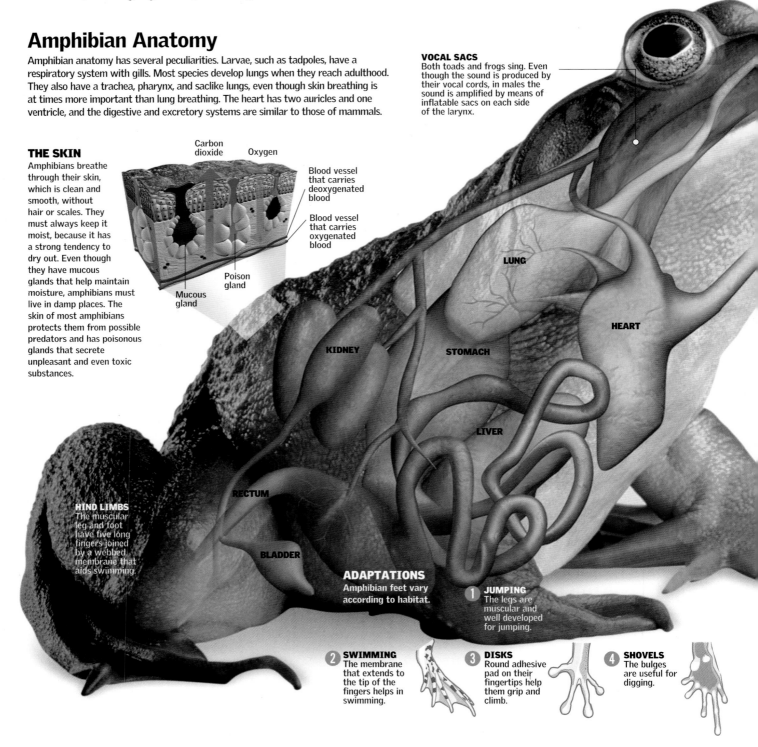

Carbon dioxide

Oxygen

Blood vessel that carries deoxygenated blood

Blood vessel that carries oxygenated blood

Poison gland

Mucous gland

LUNG

HEART

KIDNEY

STOMACH

LIVER

RECTUM

HIND LIMBS
The muscular leg and foot have five long fingers joined by a webbed membrane that aids swimming.

BLADDER

ADAPTATIONS
Amphibian feet vary according to habitat.

1 JUMPING
The legs are muscular and well developed for jumping.

2 SWIMMING
The membrane that extends to the tip of the fingers helps in swimming.

3 DISKS
Round adhesive pad on their fingertips help them grip and climb.

4 SHOVELS
The bulges are useful for digging.

DIFFERENCES BETWEEN FROGS AND TOADS

It is very common to use "frog" and "toad" as synonyms or to think that the frog is a female toad. However, frogs and toads are quite different. Toads have wrinkled skin and short legs, and they are land animals. Frogs are smaller, have webbed feet, and live in the water and in trees.

SKIN
Soft and smooth, with strong, bright colors

EYES
Frogs have horizontal pupils.

EYES
The pupil is usually horizontal, though some toads have vertical pupils.

SKIN
The skin of a toad is wrinkled, hard, rough, and dry. It is also used as leather.

COMMON TOAD
Bufo bufo.

REED FROG
Hyperolius tuberilinguis.

POSTURE
Toads are terrestrial species, slow-moving and wider than frogs. Frogs live mainly in water, which is why they have webbed toes adapted for swimming.

LEGS
are long and are adapted for jumping. Frogs have webbed toes to help with swimming.

LEGS
are shorter and wider than those of frogs and are adapted for walking.

CATCHING
Toads gulp down their prey, swallowing it whole.

SWALLOWING
Eye retraction, where the toad closes and turns its eyes inward, increases the pressure in the mouth, pushing food down the esophagus.

Nutrition

Feeding amphibians is based on plants during the larval stage, whereas in the adult stage the main food sources are arthropods (such as insects of the order Coleoptera and arachnids) and other invertebrates, such as butterfly caterpillars and earthworms.

Types of Amphibians

Amphibians are divided into three groups that are differentiated on the basis of tail and legs. Newts and salamanders have tails. They belong to the order Urodela. Frogs and toads, which have no tail except as tadpoles, belong to the Anura group. Caecilians, which have no tail or legs, are similar to worms and belong to the Apoda group.

EUROPEAN TREE FROG
Is docile and lives near buildings.

RINGED CAECILIAN
Looks like a large, thick worm.

2
APODA
Without legs.

1
ANURA
Tailless.

LEGS
Frogs and toads have four fingers on each front leg and five on each hind leg. Water frogs have webbed feet; tree frogs have adhesive disks on the tips of their fingers to hold on to vertical surfaces; and burrowing frogs have callous protuberances called tubercules on their hind legs, which they use for digging.

TIGER SALAMANDER
One of the most colorful in America.

3
URODELA
With a tail.

A Skin with Scales

Reptiles are vertebrates, meaning that they are animals with a spinal column. Their skin is hard, dry, and flaky. Like birds, most reptiles are born from eggs deposited on land. The offspring hatch fully formed without passing through a larval stage. The first reptiles appeared during the height of the Carboniferous Period in the Paleozoic Era. During the Mesozoic Era, they evolved and flourished, which is why this period is also known as the age of reptiles. Only 5 of the 23 orders that existed then have living representatives today.

SOLOMON ISLAND SKINK
Corucia zebrata.

EMBRIONARY MEMBRANES
They develop two: a protective amnion and a respiratory allantoid (or fetal vascular) membrane.

EYES
are almost always small. In diurnal animals, the pupil is rounded.

NICTITATING MEMBRANE
extends forward from the internal angle of the eye and covers it.

BLACK CAIMAN
Melanosuchus niger.

4,765
Species of lizards exist.

Habitat

Reptiles have a great capacity to adapt, since they can occupy an incredible variety of environments. They live on every continent except Antarctica, and most countries have at least one species of terrestrial reptile. They can be found in the driest and hottest deserts, as well as the steamiest, most humid rainforests. They are especially common in the tropical and subtropical regions of Africa, Asia, Australia, and the Americas, where high temperatures and a great diversity of prey allow them to thrive.

CROCODILES

Are distinguished by their usually large size. From neck to tail, their backs are covered in rows of bony plates, which can give the impression of thorns or teeth. Crocodiles appeared toward the end of the Triassic Period, and they are the closest living relatives to both dinosaurs and birds. Their hearts are divided into four chambers, their brains show a high degree of development, and the musculature of their abdomens is so developed that it resembles the gizzards of birds. The larger species are very dangerous.

OVIPAROUS

Most reptiles are oviparous (they lay eggs); however, many species of snakes and lizards are ovoviviparous (they give birth to live offspring).

THORAX AND ABDOMEN
are not separated by a diaphragm. Alligators breathe with the help of muscles on the walls of their body.

AMERICAN ALLIGATOR
Alligator mississippiensis.

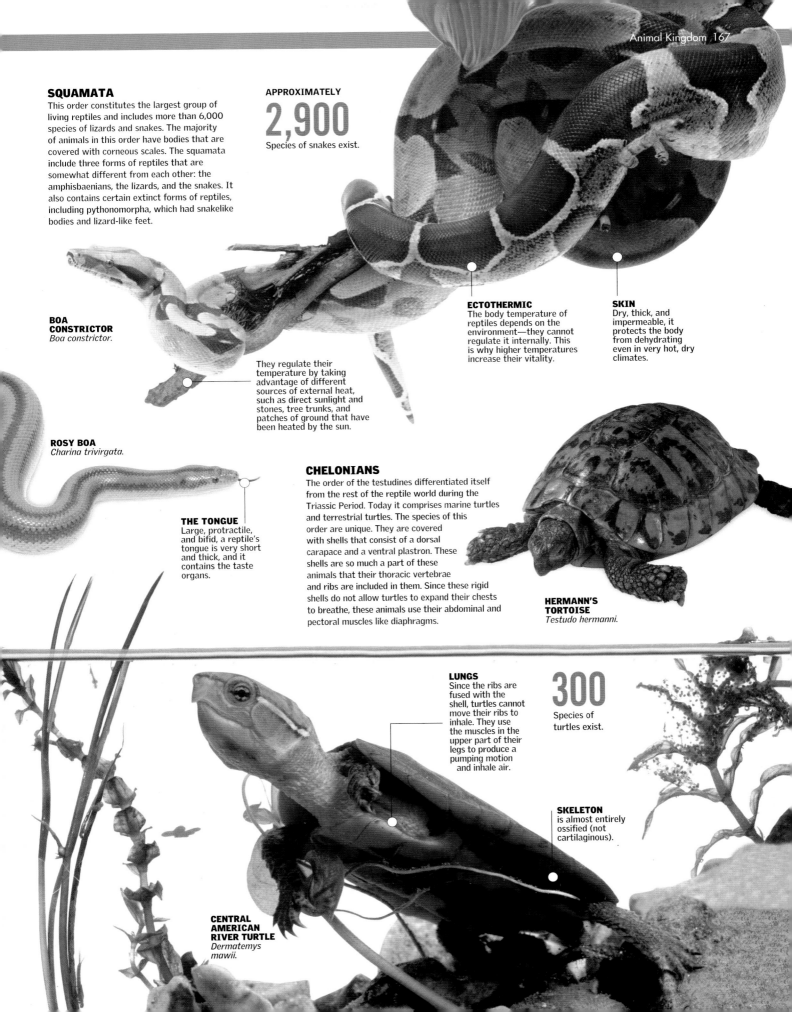

SQUAMATA

This order constitutes the largest group of living reptiles and includes more than 6,000 species of lizards and snakes. The majority of animals in this order have bodies that are covered with corneous scales. The squamata include three forms of reptiles that are somewhat different from each other: the amphisbaenians, the lizards, and the snakes. It also contains certain extinct forms of reptiles, including pythonomorpha, which had snakelike bodies and lizard-like feet.

APPROXIMATELY

2,900

Species of snakes exist.

ECTOTHERMIC
The body temperature of reptiles depends on the environment—they cannot regulate it internally. This is why higher temperatures increase their vitality.

SKIN
Dry, thick, and impermeable, it protects the body from dehydrating even in very hot, dry climates.

BOA CONSTRICTOR
Boa constrictor.

They regulate their temperature by taking advantage of different sources of external heat, such as direct sunlight and stones, tree trunks, and patches of ground that have been heated by the sun.

ROSY BOA
Charina trivirgata.

THE TONGUE
Large, protractile, and bifid, a reptile's tongue is very short and thick, and it contains the taste organs.

CHELONIANS

The order of the testudines differentiated itself from the rest of the reptile world during the Triassic Period. Today it comprises marine turtles and terrestrial turtles. The species of this order are unique. They are covered with shells that consist of a dorsal carapace and a ventral plastron. These shells are so much a part of these animals that their thoracic vertebrae and ribs are included in them. Since these rigid shells do not allow turtles to expand their chests to breathe, these animals use their abdominal and pectoral muscles like diaphragms.

HERMANN'S TORTOISE
Testudo hermanni.

LUNGS
Since the ribs are fused with the shell, turtles cannot move their ribs to inhale. They use the muscles in the upper part of their legs to produce a pumping motion and inhale air.

300

Species of turtles exist.

SKELETON
is almost entirely ossified (not cartilaginous).

CENTRAL AMERICAN RIVER TURTLE
Dermatemys mawii.

Internal Structure

Snakes are scaly reptiles with long bodies and no legs. Some are poisonous, but others are not. Like all reptiles, they have a spinal column and a skeletal structure composed of a system of vertebrae. The anatomical differences between species reveal information about their habitats and diets—climbing snakes are long and thin, burrowing snakes are shorter and thicker, and sea snakes have flat tails that they use as fins.

COLD-BLOODED
Their temperature varies according to the environment. They do not generate their own body heat.

HEART
The ventricle has an incomplete partition.

ESOPHAGUS

LUNG

LARGE INTESTINE

EMERALD TREE BOA
Corallus caninus.

TREE BRANCH
Boas can change color to imitate the branch they are curled around.

THE SPINAL COLUMN
is composed of an assembly of jointed vertebrae with prolongations that protect the nerves and arteries. The system makes them enormously flexible.

PRIMITIVE SNAKES
Boas and pythons were the first snake species to appear on Earth. Many have claws or spurs as vestiges of ancient limbs of their ancestors. They are not poisonous, but they are the largest and strongest snakes. They live in trees, and some, such as the anaconda—a South American boa—live in rivers.

33 feet (10 m)
Length of a python.

SPOTTED PYTHON
Antaresia maculosa inhabits the forests of Australia.

VERTEBRAE

- Neural arch
- Body of the vertebra
- Hemal keel

FLOATING RIBS
allow the body to increase in size.

VERTEBRA

FLOATING RIB

Range of motion of the ribs

400 vertebrae
The number a snake can have.

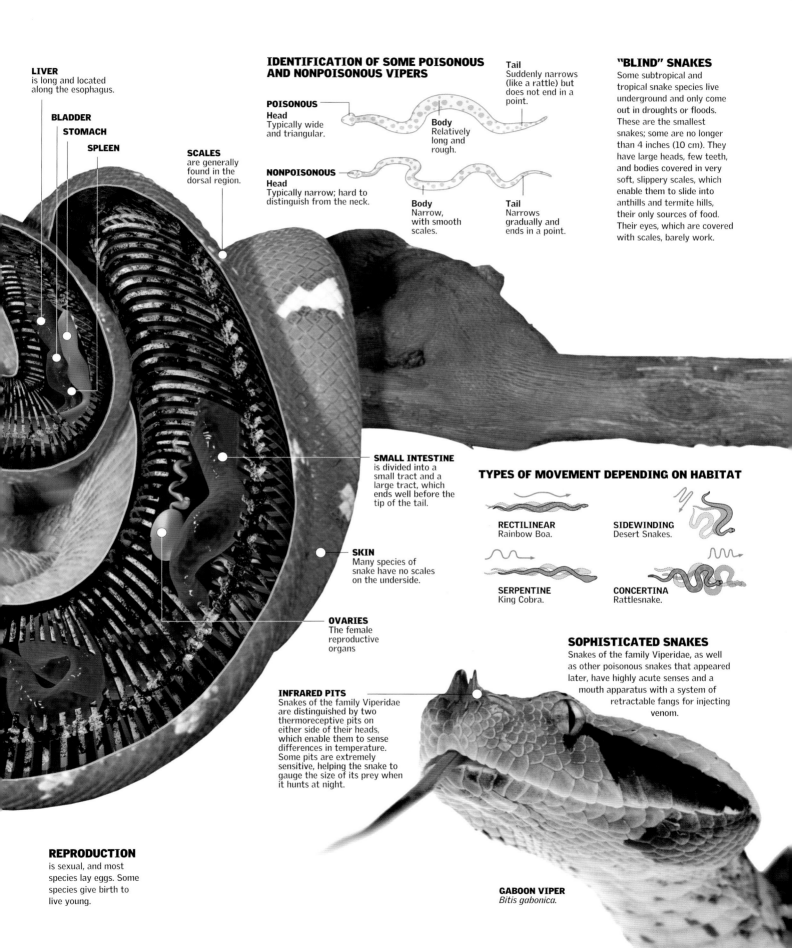

LIVER
is long and located along the esophagus.

BLADDER

STOMACH

SPLEEN

SCALES
are generally found in the dorsal region.

IDENTIFICATION OF SOME POISONOUS AND NONPOISONOUS VIPERS

POISONOUS
Head
Typically wide and triangular.

Body
Relatively long and rough.

NONPOISONOUS
Head
Typically narrow; hard to distinguish from the neck.

Body
Narrow, with smooth scales.

Tail
Suddenly narrows (like a rattle) but does not end in a point.

Tail
Narrows gradually and ends in a point.

"BLIND" SNAKES
Some subtropical and tropical snake species live underground and only come out in droughts or floods. These are the smallest snakes; some are no longer than 4 inches (10 cm). They have large heads, few teeth, and bodies covered in very soft, slippery scales, which enable them to slide into anthills and termite hills, their only sources of food. Their eyes, which are covered with scales, barely work.

SMALL INTESTINE
is divided into a small tract and a large tract, which ends well before the tip of the tail.

TYPES OF MOVEMENT DEPENDING ON HABITAT

RECTILINEAR
Rainbow Boa.

SIDEWINDING
Desert Snakes.

SERPENTINE
King Cobra.

CONCERTINA
Rattlesnake.

SKIN
Many species of snake have no scales on the underside.

OVARIES
The female reproductive organs

SOPHISTICATED SNAKES
Snakes of the family Viperidae, as well as other poisonous snakes that appeared later, have highly acute senses and a mouth apparatus with a system of retractable fangs for injecting venom.

INFRARED PITS
Snakes of the family Viperidae are distinguished by two thermoreceptive pits on either side of their heads, which enable them to sense differences in temperature. Some pits are extremely sensitive, helping the snake to gauge the size of its prey when it hunts at night.

REPRODUCTION
is sexual, and most species lay eggs. Some species give birth to live young.

GABOON VIPER
Bitis gabonica.

Jointless

The body of most mollusks is soft, extremely flexible, and without joints, yet has a large and very hard shell. Most mollusks live in the ocean, but they are also found in lakes and land environments. All modern mollusks have bilateral symmetry, one cephalopod foot with sensory organs and locomotion, a visceral mass, and a covering, called the mantle, that secretes the shell. Mollusks also have a very peculiar mouth structure called a radula.

Gastropods

These mollusks are characterized by their large ventral foot, whose wave-like motions are used to move from place to place. The group comprises snails and slugs, and they can live on land, in the ocean, and in fresh water. When these animals have a shell, it is a single spiral-shaped piece, and the extreme flexibility of the rest of the body allows the gastropod to draw itself up completely within the shell. Gastropods have eyes and one or two pairs of tentacles on their head.

BROWN GARDEN SNAIL
Helix aspersa.

DIGESTIVE GLAND

INTESTINE

LUNG

GONAD

KIDNEY

HEART

SALIVARY GLAND

ESOPHAGUS

FEMALE SEXUAL ORGAN

PROSOBRANCHIA
This mollusk subclass mainly includes marine animals. Some have mother-of-pearl on the inside of their shell, whereas others have a substance similar to porcelain.

LUNGED
Snails, land slugs, and freshwater slugs have lungs, and their lung sacs allow them to breathe oxygen in the atmosphere.

OPISTHOBRANCHIA
are sea slugs, which are characterized by having a very small shell or no shell at all.

SEA ANGEL
Candida sp.

BENDING OF THE SNAIL
In snails, bending is a very special phenomenon that moves the cavity of the mantle from the rear toward the front of the body. The visceral organs rotate 180 degrees, and the digestive tube and the nervous connections cross in a figure eight.

Gills
Nervous system
Digestive tract

Bivalves

Mollusks with a shell divided into two halves. The two parts of the shell are joined by an elastic ligament that opens the shell, adductor muscles that close the shell, and the umbo, a system of ridges that helps the shell shut together. Almost all bivalves feed on microorganisms. Some bury themselves in the wet sand, digging small tunnels that let in water and food. The tunnels can be from a fraction of an inch long to over a yard long.

SCALLOP
Pecten jacobaeus.

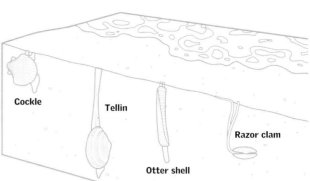

Cockle

Tellin

Razor clam

Otter shell

LAMELLIBRANCHIATA

include most bivalves. They use gills to breathe and to feed. They have no differentiated head, eyes, or extremities. They can grow up to 5 inches (13 cm) long, and they rest on the ocean floor.

GREEN MUSSEL
Perna viridis.

PROTOBRANCHIA

This class includes bivalves with a split lower foot, called a sole. Bivalves use their gills only to breathe. This subclass includes small bivalves 0.5 inch (13 mm) wide, called nutclams (*Nucula nitidosa*).

UNDER THE SAND

Many mollusks live buried under the sand in order to hide from predators and the effects of waves, wind, and sudden changes in temperature.

100,000

The number of living mollusk species; as many more have become extinct.

RADULA

Cephalopods

Cuttlefish, octopus, squid, and nautilus are called cephalopods because their extremities, or tentacles, are attached directly to their heads. These predators are adapted to life in the oceans, and they have quite complex nervous, sensory, and motion systems. Their tentacles surround their mouths, which have a radula and a powerful beak. Cephalopods can be 0.4 inch (1 cm) long to several yards long.

NAUTILOIDEA

This subclass populated the oceans of the Paleozoic and Mesozoic periods, but today only one genus–Nautilus–survives. A nautilus has an outer shell, four gills, and ten tentacles. Its shell is made from calcium, is spiral in shape, and is divided into chambers.

COLEOIDEA

Cephalopods of this class have a very small internal shell, or none at all, and only two gills. Except for the nautilus, this class includes all cephalopods alive today–octopus, cuttlefish, and squid.

COMMON CUTTLEFISH
Sepia officinalis.

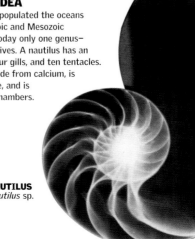

NAUTILUS
Nautilus sp.

Colorful Armor

Even though they inhabit all known environments, crustaceans are most closely identified with the aquatic environment. That environment is where they were transformed into arthropods with the most evolutionary success. Their bodies are divided into three parts: the cephalothorax, with antennae and strong mandibles; the abdomen, or pleon; and the back (telson). Some crustaceans are very small: sea lice, for instance, are no larger than one hundredth of an inch (a quarter of a millimeter). The Japanese spider crab, on the other hand, is more than 9 feet (3 m) long with outstretched legs, as it has legs in both the abdomen and the thorax in addition to two pairs of antennae.

WOODLOUSE
Armadillidium vulgare.
This invertebrate, belonging to the order Isopoda, is one of the few terrestrial crustaceans, and it is probably the one best adapted to life outside the water. When it feels threatened, it rolls itself up, leaving only its exoskeleton exposed. Even though it can reproduce and develop away from the water, it breathes through gills. The gills are found in its abdominal appendages and for this reason must be kept at specific humidity levels. That is also why the wood louse seeks dark and humid environments, such as under rocks, on dead or fallen leaves, and in fallen tree trunks.

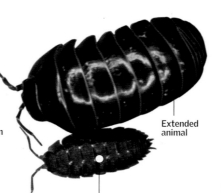

Extended animal

EXOSKELETON
Divided into independent parts.

Rolled-up animal

Antennae

Head

LEGS
This species has seven pairs of legs.

SEGMENTS
The back segments are smaller, and when they bend, they help enclose the animal completely.

Anus

Malacostraca

is the name given to the class of crustaceans that groups crabs together with sea lobsters, shrimp, wood lice, and sea lice. The term comes from Greek, and it means "soft-shelled." Sea and river crabs have 10 legs, and one pair of these legs is modified in a pincer form. Malacostraca are omnivorous and have adapted to a great variety of environments; the number of segments of their exoskeleton can vary from a minimum of 16 to more than 60.

APPENDAGES
consist of a lower region from which two segmented branches grow, one internal (endopod) and the other external (exopod).

45 pounds
(20 kg)

The Pacific spider crab can weigh up to.

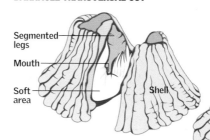

BARNACLES WITHOUT A SHELL

BARNACLE COLONY

Together Forever

At birth, barnacles *(Pollicipes cornucopia)* are microscopic larvae that travel through the sea until they reach a rocky coast. Then they attach themselves to the shore by means of a stalk, which they develop by the modification of their antennae, and then form a shell. Once they are attached, they remain in one spot for the rest of their lives, absorbing food from the water. Barnacles are edible.

BARNACLE TRANSVERSAL CUT

Segmented legs

Mouth

Soft area

Shell

Legs extended to catch food

Shell

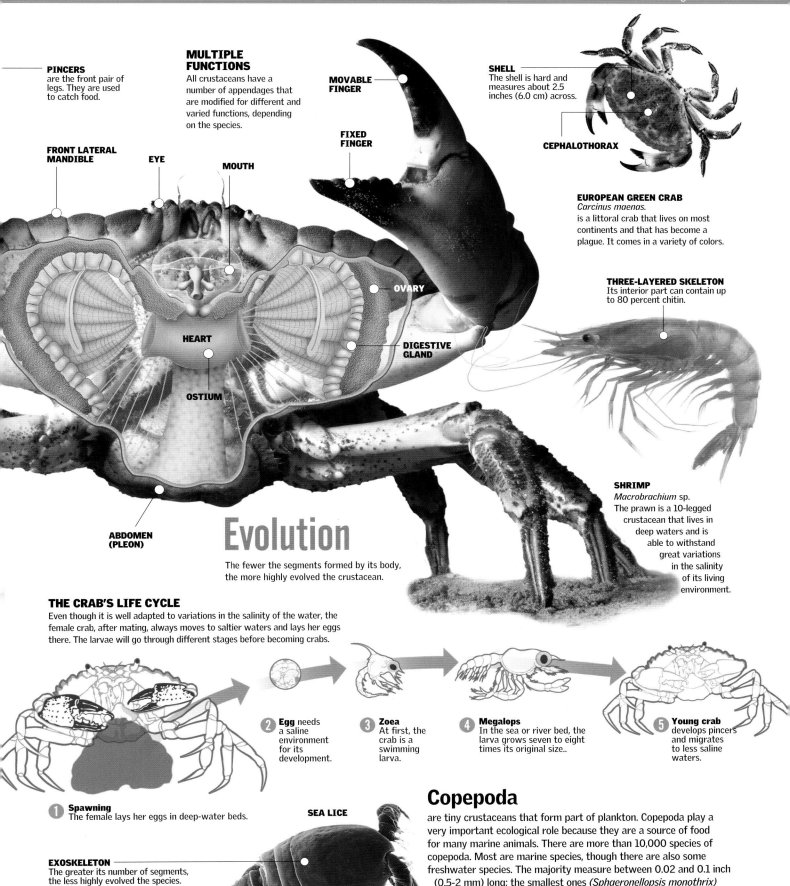

PINCERS
are the front pair of legs. They are used to catch food.

MULTIPLE FUNCTIONS
All crustaceans have a number of appendages that are modified for different and varied functions, depending on the species.

MOVABLE FINGER

FIXED FINGER

SHELL
The shell is hard and measures about 2.5 inches (6.0 cm) across.

CEPHALOTHORAX

FRONT LATERAL MANDIBLE

EYE

MOUTH

OVARY

HEART

DIGESTIVE GLAND

OSTIUM

EUROPEAN GREEN CRAB
Carcinus maenas.
is a littoral crab that lives on most continents and that has become a plague. It comes in a variety of colors.

THREE-LAYERED SKELETON
Its interior part can contain up to 80 percent chitin.

ABDOMEN (PLEON)

Evolution

The fewer the segments formed by its body, the more highly evolved the crustacean.

SHRIMP
Macrobrachium sp.
The prawn is a 10-legged crustacean that lives in deep waters and is able to withstand great variations in the salinity of its living environment.

THE CRAB'S LIFE CYCLE

Even though it is well adapted to variations in the salinity of the water, the female crab, after mating, always moves to saltier waters and lays her eggs there. The larvae will go through different stages before becoming crabs.

① **Spawning**
The female lays her eggs in deep-water beds.

② **Egg** needs a saline environment for its development.

③ **Zoea**
At first, the crab is a swimming larva.

④ **Megalops**
In the sea or river bed, the larva grows seven to eight times its original size..

⑤ **Young crab**
develops pincers and migrates to less saline waters.

SEA LICE

Copepoda

are tiny crustaceans that form part of plankton. Copepoda play a very important ecological role because they are a source of food for many marine animals. There are more than 10,000 species of copepoda. Most are marine species, though there are also some freshwater species. The majority measure between 0.02 and 0.1 inch (0.5-2 mm) long; the smallest ones (*Sphaeronellopsis monothrix*) reach only 0.004 inch (0.11 mm) in length, and the largest (*Pennella balaenopterae*) are 13 inches (32 cm) long.

EXOSKELETON
The greater its number of segments, the less highly evolved the species.

A Special Family

Arachnids make up the largest and most important class of chelicerata. Among them are spiders, scorpions, fleas, ticks, and mites. Arachnids were the first arthropods to colonize terrestrial environments. The fossil remains of scorpions are found beginning in the Silurian Period, and they show that these animals have not undergone major changes in their morphology and behavior. The most well-known arachnids are the scorpions and spiders.

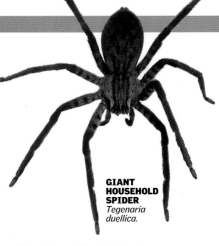

GIANT HOUSEHOLD SPIDER
Tegenaria duellica.

This spider is distinguished by its long legs in relation to its body.

The female can transport up to 30 offspring on its back.

Scorpions

Feared by people for ages, the scorpion is characterized by the fact that its chelicerae (mouth parts that in scorpions are large) and pedipalps form a pincer. The body is covered with a chitinous exoskeleton that includes the cephalothorax and abdomen.

EMPEROR SCORPION
Pandinus imperator.
Like other scorpions, it has a stinger crisscrossed by venomous glands. It measures between 5 and 7 inches (12 and 18 cm) long, though some have reached a length of 8 inches (20 cm).

The claws hold the prey and immobilize it.

PEDIPALPS
Act as sensory organs and manipulate food. Males also use them for copulation.

CHELICERAE
Move up and down. In the more primitive spiders (such as tarantulas), the chelicerae move side to side like a pincer.

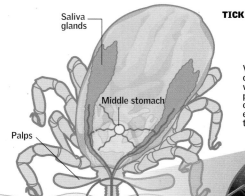

Saliva glands

TICK

Visible dorsal capitulum with dorsal projections that can be removed easily from the tick.

Middle stomach

Palps

Adhesion material

Infection

Mites and Ticks

Both are members of the Acari order. They are differentiated by their size. Mites are smaller; ticks may measure up to an inch in length (several centimeters). Mites have many diverse forms and are parasites of animals and plants. Ticks have a common life cycle of three stages: larva, nymph, and adult, during which they live off the blood of their hosts.

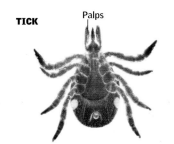

Palps

TICK

Palps

MITE

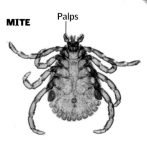

100,000

is the number of species of arachnids thought to exist in the world.

EXOSKELETON

Growth happens through molting, a process by which the spider gets rid of its old exoskeleton. In its youth the spider grows through successive moltings (up to four a year), and once it reaches adulthood, it goes through a yearly change.

1 The front edge of the shell comes off, and the tegument separates from the abdomen.

2 The spider raises and lowers its legs until the skin slips and falls.

3 It removes the old exoskeleton, and the new one hardens on contact with the air.

CEPHALOTHORAX
(PROSOMA)

ABDOMEN
(OPISTHOSOMA)

CHELICERAE

SIMPLE
EYE

HEART

INTESTINE

CLOACA

OVARIES

LUNG

VENOM
GLAND

STOMACH

SILK GLAND

GENITAL
ORIFICE

FEMUR

Spiders

Are the most common arthropods. They have the surprising property of secreting a substance that, on contact with the air, creates very fine threads that spiders skillfully manage for diverse purposes. Once a female spider mates, she deposits her eggs inside a cocoon of special silk, called an egg sack. The appearance of spiders is unmistakable: the two main sections of the body, the thorax (also called a prosoma) and the abdomen (also called an opisthosoma), are united by a narrow stalk (the pedicel). Spiders have four pairs of eyes, whose distinctive size and placement help characterize different families of spiders. Their chelicerae end in fangs that carry conduits from venom glands. Spiders kill their prey by using their chelicerae to apply venom.

PATELLA

TIBIA

PEDIPALPS
The terminal pedipalp forms a copulating organ through which the male inseminates the female.

WALKING LEGS
The spider has four pairs of legs for walking. The hairs help it to recognize terrain.

METATARSUS

Amblypygi

Small arachnids measure between 0.2 and 2 inches (0.4 and 4.5 cm). The chelicerae are not as large, though the pedipalps are strong and are used to capture prey. The first pair of legs are modified touch-and-sensing appendages, whereas the last three take care of movement. Because of a spider's flattened body, its walk is similar to that of the crab.

TARSUS

12 inches
(30 cm)

A spider can measure with its legs spread out.

**PEDIPALP
ARACHNID**
*Phryna
grossetaitai.*

BEAUTIFUL FLOWERS
Red water lilies on a lake in
Norshingali, on the outskirts of
Dhaka, Bangladesh.

CHAPTER 6

VEGETAL
KINGDOM

Plants provide us with food, medicines, wood, resins, and oxygen, among other essential products, thanks to photosynthesis. One of Earth's fundamental building blocks, without plants humans would not be able to live.

Conquest of Land

The movement of plants from shallow water onto land is associated with a series of evolutionary events. Certain changes in the genetic makeup of plants enabled them to face the new and extreme conditions found on the Earth's surface. Although land habitats offered plants direct exposure to sunlight, they also presented the problem of transpiration and the loss of water that it produces. This difficulty had to be overcome before plants could spread over land.

Vital Changes

Roots are among the most important adaptations for plants' success in land habitats. Root systems anchor the plant in the substrate and serve as a pathway for water and mineral nutrients to enter it. Besides roots, the development of a cuticle (skin membrane) to cover the entire plant's surface was crucial. Cells in the epidermis produce this waterproof membrane, which helps the plant tolerate the heat generated by sunlight and the wear and loss of water caused by the wind. This protection is interrupted by pores, which allow for gas exchange.

GREEN REVOLUTION

Leaves are the main organs for photosynthesis in land plants. After plants appeared on land more than 440 million years ago, the amount of photosynthesis taking place gradually increased. This increase is believed to be one of the reasons the concentration of carbon dioxide in the atmosphere decreased. As a result, the Earth's average temperature also decreased.

50,000

Species of fungus live alongside land-dwelling plants.

MALE FERN
Dryopteris filix-mas.
These vascular plants need liquid water to reproduce.

MOSS
Sphagnum sp.
Bryophytes are the simplest of all land plants.

EPIPHYTES

grow on plants or on some other supporting surface. Their anatomy includes secondary adaptations that enable them to live without being in contact with the soil.

GRASSES

take advantage of long hours of summer daylight to grow and reproduce. Their stalks do not have reinforcing tissues that would enable them to remain erect.

STEMLESS SOW THISTLE
Sonchus acaulis.
These plants lack a stem.

SWEET VIOLET
Viola odorata.
This plant's spring flowers have a pleasant scent.

GIANTS

Trees are distinguished by their woody trunks. As a tree grows from a tender shoot, it develops a tissue that gives it strength, enabling it to grow over 330 feet (100 m) tall. Trees are found in the principal terrestrial ecosystems.

CHESTNUTS
Castanea sp.

WALNUTS
Juglans sp.

BEECHES
Fagus sp.

MAPLES
Acer sp.

OAKS
Quercus sp.

LINDENS
Tilia sp.

360 feet
(110 m)

The height reached by some sequoia sempervirens trees.

Anatomy of a Tree

The oak tree is the undisputed king of the Western world. It is known for its lobed leaves and the large cap of its acorn, a nut found on all trees of the genus Quercus. The tree's main trunk grows upward and branches out toward the top. Oaks are a large group, containing many types of deciduous trees. Under optimal conditions oaks can grow to a height of more than 130 feet (40 m) and live an average of 600 years.

Leaves

are arranged one leaf to a stem on alternating sides of the twig. They have rounded lobes on either side of the main vein.

The leaves absorb CO_2 and produce sugars by means of photosynthesis.

SUMMER
The leaves undertake photosynthesis, and the rest of the tree uses the sugars it produces.

AUTUMN
The cells at the end of each leaf stem weaken.

WINTER
The leaf falls away, and the tree remains dormant.

SPRING
New leaves begin to replace the old ones.

BARK

GROWTH RINGS

Seasons

SPRING
The cycle begins as the first leaves appear.

SUMMER
The oak blossoms. It increases in height, and its trunk grows thicker.

AUTUMN
Low temperatures weaken the branches.

WINTER
The leaves fall away; the tree is dormant until spring.

Trunk

The trunk is strong and grows straight upward. The top of the tree widens with branches, which may be twisted, knotted, or bent.

Seeds

Some species have sweet-tasting seeds; others are bitter.

BEGINNINGS
In its first year of life an oak tree's roots can grow nearly 5 feet (1.5 m).

Transpiration (the loss of water vapor) in the leaves pulls the xylem sap upward.

Buds

are formed by protective scales that fall off in the spring. They grow into new leaves and branches.

Woodpeckers drill holes in the tree with their beaks as they look for insects.

Flowers

The tree produces hanging male flowers, whereas female flowers are hidden among the leaves.

Oak-Tree Products

The bark is rich in tannin, which is used in curing leather and as an astringent. The wood is strong and resists rotting.

ENERGY SOURCE

The chlorophyll traps energy from sunlight and uses it to convert water and carbon dioxide into food.

The xylem transports water and minerals from the roots to the rest of the tree.

The phloem transports sugars from the leaves to the rest of the tree.

Acorns

have dark stripes along their length. Their caps have flat scales.

Remains of the Carpel (female reproductive part).

Achene: A hard seed that does not split open at maturity.

SURFACE

Mosses use the bark of oak trees as a source of moisture.

600 years

The average life span of an Oak.

ROOTS
grow sideways to form a deep, broad root system.

Absorption of Water and Minerals

Feeding on Light

An important characteristic of plants is their ability to use sunlight and the carbon dioxide in the air to manufacture their own complex nutrients. This process, called photosynthesis, takes place in chloroplasts, cellular components that contain the necessary enzyme machinery to transform solar energy into chemical energy. Each plant cell can have between 20 and 100 oval-shaped chloroplasts. Chloroplasts can reproduce themselves, suggesting that they were once autonomous organisms that established a symbiosis, which produced the first plant cell.

Why Green?

Leaves absorb energy from visible light, which consists of different colors. The leaves reflect only the green light.

ALGAE
perform photosynthesis underwater. Together with water plants, they provide most of the atmosphere's oxygen.

LEAVES
are made of several types of plant tissues. Some serve as a support, and some serve as filler material.

O_2
is released by plants into the Earth's atmosphere

PLANT CELLS
have three traits that differentiate them from animal cells: cell walls (which are made up of 40 percent cellulose), a large vacuole containing water and trace mineral elements, and chloroplasts containing chlorophyll. Like an animal cell, a plant cell has a nucleus.

WATER
Photosynthesis requires a constant supply of water, which reaches the leaves through the plant's roots and stem.

CHLOROPHYLL
is the most abundant pigment in leaves.

CELL MEMBRANE

CELL WALL

PLANT TISSUES
The relative stiffness of plant cells is provided by cellulose, the polysaccharide formed by the plant's cell walls. This substance is made of thousands of glucose units, and it is very difficult to hydrolyze (break down in water).

CARBON DIOXIDE
is absorbed by plant cells to form sugars by means of photosynthesis.

OXYGEN
is a by-product of photosynthesis. It exits the surface of the leaves through their stoma (two-celled pores).

VACUOLE
provides water and pressure and gives the cell consistency.

Stages of the Process

Photosynthesis takes place in two stages. The first, called photosystem II, depends directly on the amount of light received, which causes the chlorophyll to release electrons. The resulting gaps are filled by electrons of water, which breaks down and releases oxygen and ionized hydrogen (2H+).

1 ATP (Adenosine Triphosphate) formation is powered by the movement of electrons into receptor molecules in a chain of oxidation and reduction reactions.

2 In photosystem I light energy is absorbed, sending electrons into other receptors and making NADPH (Nicotinamide Adenine Dinucleotide Phosphate Hydrogen) out of NADP+.

3 The ATP and NADPH obtained are the net gain of the system, in addition to oxygen. Two water molecules are split apart in the process, but one is regenerated when the ATP is formed.

PHOTOSYSTEM I

NADPH

2H+

PHOTOSYSTEM II

PROTEIN

NADP+ REDUCTASE

FLOW OF ELECTRONS

H_2O

O_2

2H+

4 In photosystem I ATP is also generated from ADP (Adenosine Diphosphate) because of the surplus flow of free electrons.

THYLAKOIDS
Sacs that contain chlorophyll molecules. Inside them ADP is converted into ATP as a product of the light-dependent phase of photosynthesis. Stacked thylakoids form a structure called a grana.

THYLAKOID MEMBRANE

ADP + P

H+

ATP

GRANA

NUCLEOLE

NUCLEUS

CO_2

P + ADP

ATP

H + NADP+

NADPH

Calvin Cycle

END PRODUCTS enable the plant to generate carbohydrates, fatty acids, and amino acids.

CHLOROPLAST
The part of the cell where both phases of photosynthesis take place. It also contains enzymes.

STROMA
is the watery space inside the chloroplast.

The Dark Phase

This phase, so called because it does not directly depend on light, takes place inside the stroma of the chloroplast. Energy in the form of ATP and NADPH, which was produced in the light-dependent phase, is used to fix carbon dioxide as organic carbon through a process called the Calvin cycle. This cycle consists of chemical reactions that produce phosphoacylglycerides, which the plant cell uses to synthesize nutrients.

Carbon
The building block of organic material.

Aquatic Plants

These plants are especially adapted for living in ponds, streams, lakes, and rivers—places where other land plants cannot grow. Although aquatic plants belong to many different families, they have similar adaptations and are therefore an example of adaptive convergence. They include submerged plants and floating plants; plants that may or may not be rooted at the bottom; amphibious plants, which have leaves both above and below the water's surface; and heliophilic plants, which have only their roots underwater.

A Vital Role

Aquatic plants play an important role in the ecosystem not only for crustaceans, insects, and worms but also for fish, birds, and mammals because they are an important source of food and shelter for these categories of animals. Aquatic plants also play a major role in converting solar energy into the organic materials upon which many living things depend.

ROOTED PLANTS WITH FLOATING LEAVES

Such plants are often found in standing or slow-moving water. They have fixed rhizomes and petiolate leaves (leaves with a stalk that connects to a stem) that float on the surface of the water. Some of the plants have submerged leaves, some have floating leaves, and some have leaves outside the water, with each type having a different shape. In the case of floating leaves the properties of the upper surface are different from those of the lower surface, which is in contact with the water.

PARROT FEATHER
Myriophyllum aquaticum
This plant is native to temperate, subtropical, and tropical regions, and it is highly effective at oxygenating water.

TROPICAL WATER LILY
Victoria cruciana
It grows in deep, calm waters. Its leaves can measure up to 7 feet (2 m) across.

FLOATING LEAVES
The rhizomes are fixed, the leaves grow on long stalks, and the leaf surface floats on the water.

Upper Epidermis

Parenchyma
Aerenchyma

Lower Epidermis

Conduction Bundle

Air Chamber

YELLOW FLOATING HEART
Nymphoides peltata
It produces small creased yellow flowers all summer long.

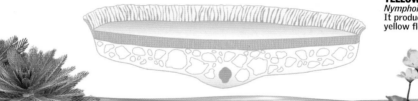

ROOTED UNDERWATER PLANTS

The entire plant is submerged. The small root system serves only to anchor the plant since the stem can directly absorb water, carbon dioxide, and minerals. These plants are often found in flowing water. The submerged stems have no system of support—the water holds up the plant.

SAGO PONDWEED
Potamogeton densus
This water plant can be found in shallow depressions of clear-flowing streams.

HORNWORT
Ceratophyllum sp.
This plant has an abundance of fine leaves that form a conelike structure on each stem.

They produce and release oxygen as a result of photosynthesis.

Aquatic but Modern

The evolutionary history of plants began in water environments. They later conquered land by means of structures such as roots. Modern aquatic plants are not a primitive group, however. On the contrary, they have returned to the water environment by acquiring highly specialized organs and tissues. For example, some tissues have air pockets that enable the plant to float.

AERENCHYMA

is always found in floating organisms. This tissue has an extensive system of intercellular spaces through which gases are diffused.

Aerenchyma

Epidermis

Air Chamber

Submerged stems have no support system because the water holds up the plant. Their limiting factor is oxygen availability, so the aerenchyma helps make this substance available to the plant.

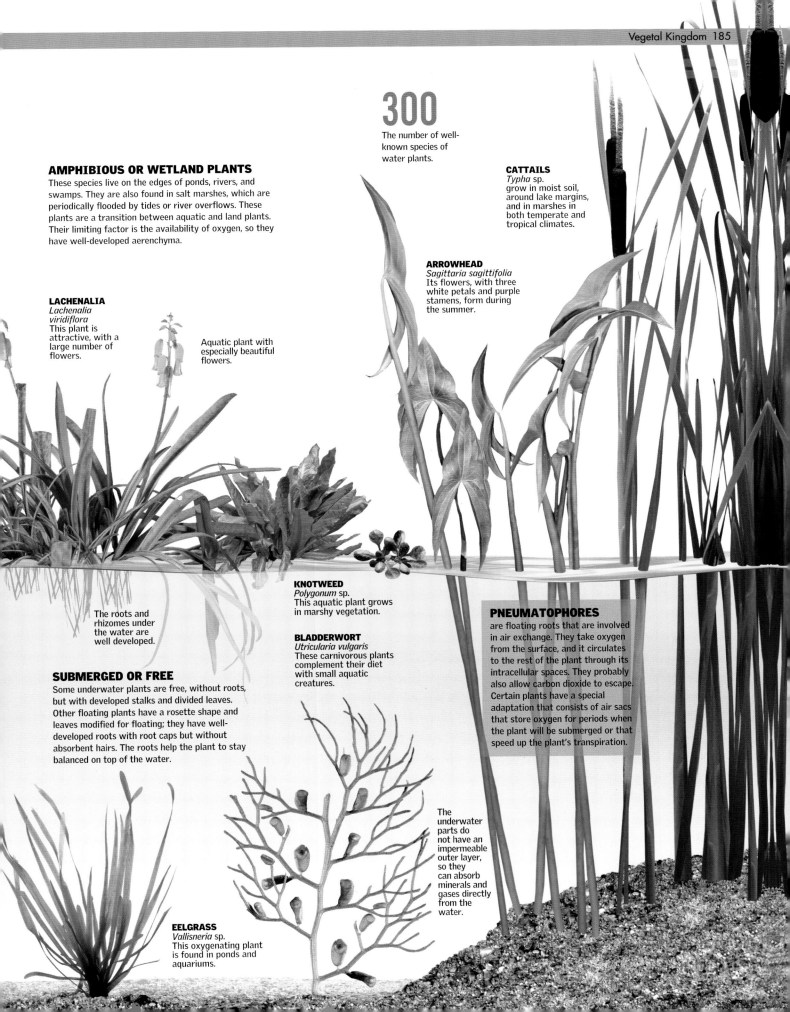

300
The number of well-known species of water plants.

AMPHIBIOUS OR WETLAND PLANTS
These species live on the edges of ponds, rivers, and swamps. They are also found in salt marshes, which are periodically flooded by tides or river overflows. These plants are a transition between aquatic and land plants. Their limiting factor is the availability of oxygen, so they have well-developed aerenchyma.

CATTAILS
Typha sp.
grow in moist soil, around lake margins, and in marshes in both temperate and tropical climates.

ARROWHEAD
Sagittaria sagittifolia
Its flowers, with three white petals and purple stamens, form during the summer.

LACHENALIA
Lachenalia viridiflora
This plant is attractive, with a large number of flowers.

Aquatic plant with especially beautiful flowers.

The roots and rhizomes under the water are well developed.

KNOTWEED
Polygonum sp.
This aquatic plant grows in marshy vegetation.

PNEUMATOPHORES
are floating roots that are involved in air exchange. They take oxygen from the surface, and it circulates to the rest of the plant through its intracellular spaces. They probably also allow carbon dioxide to escape. Certain plants have a special adaptation that consists of air sacs that store oxygen for periods when the plant will be submerged or that speed up the plant's transpiration.

BLADDERWORT
Utricularia vulgaris
These carnivorous plants complement their diet with small aquatic creatures.

SUBMERGED OR FREE
Some underwater plants are free, without roots, but with developed stalks and divided leaves. Other floating plants have a rosette shape and leaves modified for floating; they have well-developed roots with root caps but without absorbent hairs. The roots help the plant to stay balanced on top of the water.

The underwater parts do not have an impermeable outer layer, so they can absorb minerals and gases directly from the water.

EELGRASS
Vallisneria sp.
This oxygenating plant is found in ponds and aquariums.

Seeds, To and Fro

Reproduction from seeds is the most prominent evolutionary advantage in plants' conquest of the terrestrial environment. The seed shelters the embryo of the future plant with protective walls. The embryo is accompanied by tissues that provide enough nutrients for it to begin to develop. Optimal temperature and an appropriate quantity of water and air are the factors that stimulate the seed to awaken to a marvelous cycle of development and growth that will culminate in the generation of new seeds.

1 **AWAKENING OF THE SEED**
Seeds, such as those of the field, or corn, poppy *(Papaver rhoeas)*, leave their latent stage when they hydrate and receive enough light and air. Their protective coverings open and the embryo grows thanks to the energy provided by its cotyledons, or seed leaves.

2 **TROPISM**
Because of gravity, amyloplasts are always located in the lower part of cells. They produce a stimulus that encourages the root to grow toward the earth, a process called geotropism.

Cell multiplication allows the stem to grow.

PLUMULE
The bud of a plant embryo that will produce the first shoot

COTYLEDON
The first embryo leaf. It provides the energy needed for growth.

ABSORBENT HAIRS
These organs begin to develop in the radicle. They help the seed absorb water from the soil.

HARD COVER
Called the testa, it can appear in very different forms.

The testa protects the embryo and the cotyledons during the seed's latent stage.

RADICLE
The embryo root that will produce the main root of the plant

Enzymes — Nutrients

ENDOSPERM

Gibberellin — Embryo

Seed Cover

WATER
is responsible for breaking open seed covers because the hydrated tissues exert pressure on the interior of the seed.

NUTRIENTS
The radicle is in charge of collecting water and nutrients present in the soil.

Gibberellins

are plant hormones that, during the first stages of germination following water absorption, are distributed through the endosperm. Their presence promotes the production of enzymes that hydrolyze starches, lipids, and proteins to turn them into sugars, fatty acids, and amino acids, respectively. These substances provide nutrition to the embryo and later to the seedling.

Autumn

The time of the year in wich the seed of *Papaver rhoeas* germinates.

3 **GROWTH**
The seedling grows and breaks through the surface. This causes the plant to be exposed to light so it can begin to carry out photosynthesis. It thus begins to manufacture its own nutrients to replace those provided by the cotyledons.

4 **VEGETATIVE GROWTH**
The first true leaves unfold above the cotyledons, and the stem elongates from formative tissue called the meristem, located at the apex of the plant. Continued growth will lead to the formation of an adult plant, which will develop its own reproductive structures.

FLOWERING
Internal and external changes stimulate the apical bud to develop a flower.

APICAL GROWTH
Light stimulates the multiplication of cells in the apex of the stem.

The cotyledon is carried by the vertical growth of the stem.

Cotyledons can remain under the soil or, as in this case, grow above the ground.

HYPOCOTYL
The first part of the stem that emerges and develops in the young plant

TOTIPOTENCY
Characteristic of the vegetative apex cells

FIRST TRUE LEAVES

SESSILE LEAVES
The upper leaves have no petiole.

5 **PRODUCTION OF THE FLOWER'S PARTS**
The apical bud begins to produce fertile flower structures (gynoecium and androecium) and sterile structures (petals and sepals). The flower bud forms.

CONDUCTION
The stem carries water and minerals from the root to the leaves, while taking manufactured substances in the opposite direction.

0.4 inch
(1 cm)
Is the maximun height it can grow in one day.

ALTERNATE LEAVES

PRIMARY ROOT
It anchors itself to the ground and branches out to support the plant in the substrate.

SECONDARY ROOTS

The root has many fine hairs that create a large surface area for water absorption.

THE FIRST 20 DAYS OF A FIELD POPPY

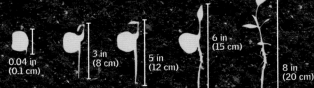

0.04 in
(0.1 cm)

3 in
(8 cm)

5 in
(12 cm)

6 in
(15 cm)

8 in
(20 cm)

20 inches
(50 cm)
The typical height of an adult field poppy plant

Under the Earth

The root is a plant organ that is usually found under the soil. It has positive geotropism; its main functions are absorbing water and inorganic nutrients and attaching the plant to the ground. The root is essential for identifying the particular characteristics of a plant. The anatomical structure of a root can vary, but, because it does not have leaves or nodes, it will always be simpler than that of a stem.

Types of Roots

Roots differ, depending on their origin. The primary root originates in the radicle of the embryo. An adventitious root is one that originates in any other organ of the plant. Roots are also subdivided according to their morphology.

TAPROOT
A taproot grows downward and has lateral secondary roots that are not well developed.

BRANCHED
The main root is divided, creating other secondary roots.

FIBROUS
The root system is formed by a group of roots of similar diameter.

TUBEROUS
Fibrous in structure, some of the roots thicken to store food for the plant.

NAPIFORM
The taproot thickens with stored food and tapers abruptly near its tip.

TABULAR
Tabular roots form at the base of a trunk and create a supporting buttress.

GEOTROPISM

Geotropism, or gravitotropism, is the growth of a plant or parts of a plant in a particular direction because of the stimulus of gravity. The force of gravity orients the stems and their leaves to grow upward (negative geotropism), whereas the roots grow downward (positive geotropism).

Monocotyledons

These plants have embryos with only one cotyledon. Their embryonic root generally has a relatively short life and is replaced by adventitious roots that grow from the stem.

CELLULAR DIVISION

Through the process of cell division a cell divides into two cells, each with its own nucleus. The new cells elongate, allowing the root to grow in thickness and length.

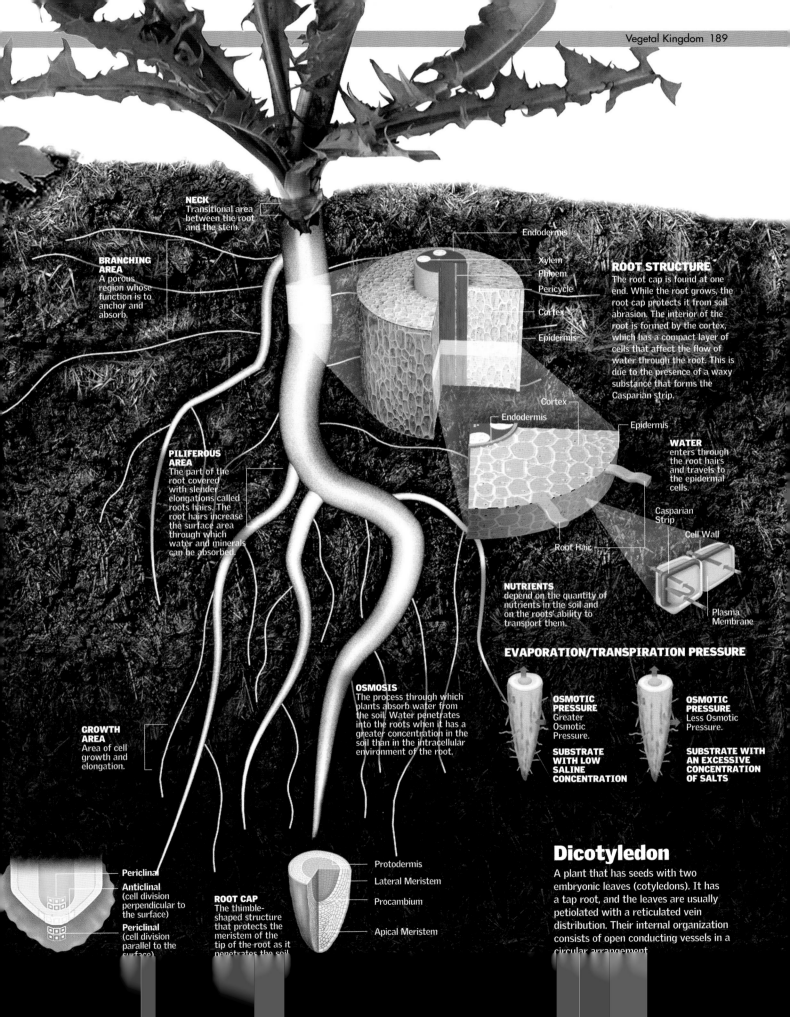

NECK
Transitional area between the root and the stem.

BRANCHING AREA
A porous region whose function is to anchor and absorb.

PILIFEROUS AREA
The part of the root covered with slender elongations called roots hairs. The root hairs increase the surface area through which water and minerals can be absorbed.

GROWTH AREA
Area of cell growth and elongation.

Periclinal

Anticlinal (cell division perpendicular to the surface)

Periclinal (cell division parallel to the surface)

ROOT CAP
The thimble-shaped structure that protects the meristem of the tip of the root as it penetrates the soil.

Protodermis
Lateral Meristem
Procambium
Apical Meristem

OSMOSIS
The process through which plants absorb water from the soil. Water penetrates into the roots when it has a greater concentration in the soil than in the intracellular environment of the root.

Endodermis
Xylem
Phloem
Pericycle
Cortex
Epidermis

ROOT STRUCTURE
The root cap is found at one end. While the root grows, the root cap protects it from soil abrasion. The interior of the root is formed by the cortex, which has a compact layer of cells that affect the flow of water through the root. This is due to the presence of a waxy substance that forms the Casparian strip.

Cortex
Endodermis
Epidermis

WATER
enters through the root hairs and travels to the epidermal cells.

Casparian Strip
Cell Wall
Root Hair
Plasma Membrane

NUTRIENTS
depend on the quantity of nutrients in the soil and on the roots' ability to transport them.

EVAPORATION/TRANSPIRATION PRESSURE

OSMOTIC PRESSURE
Greater Osmotic Pressure.

SUBSTRATE WITH LOW SALINE CONCENTRATION

OSMOTIC PRESSURE
Less Osmotic Pressure.

SUBSTRATE WITH AN EXCESSIVE CONCENTRATION OF SALTS

Dicotyledon

A plant that has seeds with two embryonic leaves (cotyledons). It has a tap root, and the leaves are usually petiolated with a reticulated vein distribution. Their internal organization consists of open conducting vessels in a circular arrangement.

Stems: More Than a Support

Stems, which occur in a variety of shapes and colors, support a plant's leaves and flowers. They keep it from breaking apart in the wind, and they determine its height. In addition, stems are also responsible for distributing the water and minerals absorbed by a plant's roots. Stems contain conducting vessels through which water and nutrients circulate. In trees and bushes, stems are woody for better support.

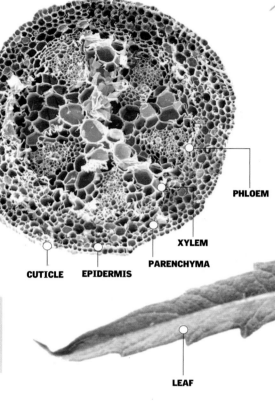

CROSS-SECTION OF A NEW STEM

PHLOEM

XYLEM

PARENCHYMA

CUTICLE   EPIDERMIS

LEAF

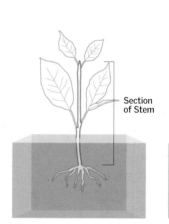

Section of Stem

IN THE AIR
Stems are usually branched, as seen in trees and bushes.

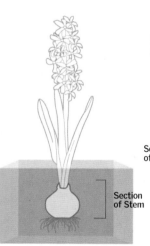

Section of Stem

Section of Stem

IN THE GROUND
Certain types of stems have unusual characteristics.

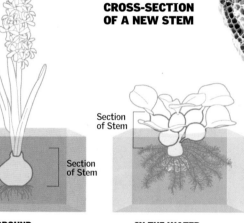

Section of Stem

IN THE WATER
The stem of an aquatic plant can lie underwater.

Development of Stems in Different Mediums

Stems have widely varying sizes and shapes that reflect different adaptations to the environment. Palm trees and wheat are two good examples that show how different mediums can modify the stem through evolution. Palm trees are the tallest non-woody plants. They grow tall because they must compete with many other plants for sunlight. In contrast, wheat is typical of areas with a cold climate and a short growing season. It develops a relatively short stem. This enables it to survive the physical assault of the dry wind and the loss of leaves.

SPROUTS
grow from the eyes.

TUBER
An underground stem composed mainly of parenchymatic cells filled with starch. The potato's small depressions are actually axillary eyes. In an onion, another example of a plant with an underground stem, starch accumulates not in tubers but in thick leaves that grow around the stem.

ARTICHOKE THISTLE
Cynara cardunculus.

COMMON POTATO
Solanum tuberosum.

AXILLARY EYES
are grouped in a spiral pattern along the potato.

Circulation

Because the stem is the link between the roots, which absorb water and minerals, and the leaves, which produce food, the stem's veined tissues are connected to the roots and leaves. It functions as a transport system for interchanging substances. The stem and its branches hold the leaves up to receive light and support the plant's flowers and fruit. Some stems have cells with chlorophyll that carry out photosynthesis; others have specialized cells for storing starch and other nutrients.

MOVEMENT THROUGH THE STEM

In plants, sugar and other organic molecules are transported through the phloem, which moves the sap. The molecules are transported through sieve tubes.

GLUCOSE
Sugar reduces the osmotic pressure in the sieve tubes.

WATER AND SALTS
are absorbed by the roots and then transported and distributed by the xylem in the stem.

CORE

XYLEM

CAMBIUM

PHLOEM

AXILLA
The joint between the main stem and a leaf stem

NODE
A place where shoots grow from the stem

INTERNODE
The part of the stem between two nodes

CROSS-SECTION OF STEM

XYLEM VESSEL

HEARTWOOD

SAPWOOD

COMPANION CELL

SIEVE PLATE

SIEVE-TUBE ELEMENT

SIEVE TUBE

PRIMARY PHLOEM

SECONDARY PHLOEM

INNER BARK

Energy Manufacturers

The main function of leaves is to carry out photosynthesis. Their shape is specialized to capture light energy and transform it into chemical energy. Their thinness minimizes their volume and maximizes their surface area that is exposed to the Sun. However, there are a great many variations on this basic theme, which have evolved in association with different types of weather conditions.

EDGES (MARGINS)
Species are distinguished by a wide variety of edges: smooth, jagged, and wavy.

VEINS
Flowering plants (division Angiosperma) are often distinguished by the type of veins they have: parallel veins in monocots and branching veins in dicots.

PRIMARY VEINS
The products of photosynthesis circulate through the veins from the leaves to the rest of the body.

RACHIS

ACER SP.
This genus includes trees and bushes easily distinguishable by their opposite and lobed leaves.

LEAF STEM (PETIOLE)

LEAF SURFACE
Colorful, usually green, with darker shades on the upper, or adaxial, side. The veins can be readily seen.

Simple Leaves

In most monocotyledon plants the leaf is undivided. In some cases it may have lobes or notches in its side, but these divisions do not reach all the way to the primary vein of the leaf.

Compound Leaves

When the leaf is divided from the primary vein, it forms separate leaflets. A compound leaf is called palmate when the leaflets are arranged like the fingers on a hand and pinnate when they grow from the sides of the leaf stem like the barbs of a feather.

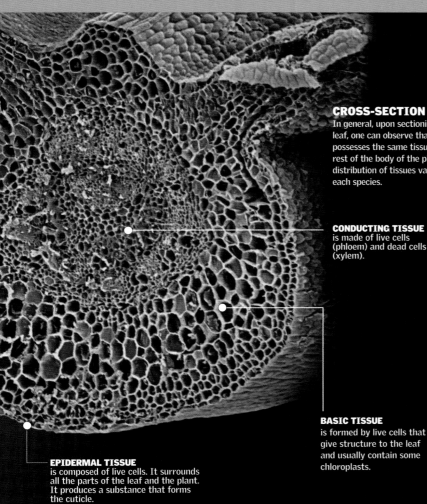

CROSS-SECTION
In general, upon sectioning a leaf, one can observe that it possesses the same tissues as the rest of the body of the plant. The distribution of tissues varies with each species.

CONDUCTING TISSUE
is made of live cells (phloem) and dead cells (xylem).

BASIC TISSUE
is formed by live cells that give structure to the leaf and usually contain some chloroplasts.

EPIDERMAL TISSUE
is composed of live cells. It surrounds all the parts of the leaf and the plant. It produces a substance that forms the cuticle.

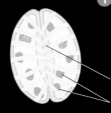

1 The stomatic apparatus is closed. No air can enter or leave the leaf. This prevents excessive transpiration, which could damage the plant.

Thickened cell walls in the area of the pore.

Cellulose Microfibers

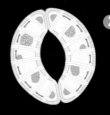

2 The stomatic apparatus is open. The stomatic cells are swollen. As tension increases, the cellular form is modified and is able to exchange gases.

PLANTS AND THE ENVIRONMENT
The flow of carbon dioxide and water vapor between the plant and the environment is essential for the photosynthetic process. This exchange can be affected by internal or external factors, such as changes in light, temperature, or humidity. In response to these stimuli the stomas can open or close.

Change and Its Advantages
Conifers possess an interesting modification in their leaves. In these gymnosperms evolution directed the abrupt reduction of surface foliage area. This gave them an adaptive advantage over plants whose leaves have a large surface area: less resistance to wind and less transpiration in arid climates. In addition, they are able to avoid the excessive weight that would result from the accumulation of snow on large leaves.

VASCULAR BUNDLE
Formed by phloem and xylem

RESIN
functions to prevent freezing. It circulates through the resin ducts.

EPIDERMIS
Cells with thick walls and a thick cuticle

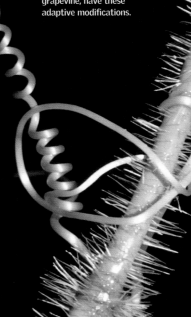

TENDRILS
The leaves of climbing plants, such as the grapevine, have these adaptive modifications.

CONIFERS
Needle-shaped leaves are characteristic of conifers. They are usually oval or triangular. A hypodermis, which is enclosed by the epidermis, is broken only in the stomas.

Functional Beauty

Flowers are not simply beautiful objects; they are also the place where the reproductive organs of angiosperms are located. Many are hermaphroditic, meaning that they contain both the male reproductive apparatus (the androecium) and the female (the gynoecium). The process of pollination is carried out through external agents, such as insects, birds, wind, and water. Following fertilization, flowers produce seeds in their ovaries. The floral parts are arranged in circular or spiral patterns.

GYNOECIUM
The female reproductive system. It is formed by carpels and includes the ovary, ovules, style, and stigma.

STIGMA
It can be simple or divided. It secretes a sticky liquid that captures the pollen. Some are also covered with hair.

Classification

Plants with flowers are classified as dicotyledons or monocotyledons. The first group has seeds with two cotyledons, and the second has seeds with only one. Each represents a different evolutionary line. They are differentiated by the structure of their organs. The cotyledon contains nutrients that the embryo utilizes during its growth until its true leaves appear. When a seed germinates, the first thing that appears is the root. In monocotyledons the stem and the radicle are protected by a membrane; the dicotyledons lack this protection, and the stem pushes itself through the soil.

DICOTYLEDONS

In this class of plants each whorl of the flower is arranged in groups of four or five parts. In dicotyledons the sepal is small and green, the petals are large and colorful, and the leaves are wide. The vascular ducts are cylindrical.

FLORAL DIAGRAM

MONOCOTYLEDONS

Each whorl of these flowers contains three parts, and their sepals and petals are generally not differentiated from one another. The majority are herbaceous plants with scattered vascular conduits. They are the most evolved species of angiosperms.

FLORAL DIAGRAM

OVARY
The ovary is found in the receptacle at the base of the gynoecium, inside the carpels. The pollen tube extends into the ovary and penetrates the ovule.

CARPEL
The carpel consists of modified leaves that together form the gynoecium. It contains a stigma, a style, and an ovary. Ovules are produced in the ovary.

LEAVES
In dicotyledons, leaves have various forms, and they contain a network of veins that connect with a primary vein.

ROOT
In dicotyledons the main root penetrates the ground vertically as a prolongation of the stem, and secondary roots extend from it horizontally. It can be very deep and long-lived.

LEAVES
Plants with only one cotyledon have large and narrow leaves, with parallel veins and no petiole.

ANDROECIUM
The male reproductive system. It is formed by a group of stamens, each of which consists of an anther supported by a filament. The base may contain glands that produce nectar.

ANTHER
A sac where grains of pollen (the male gametes) are produced.

FILAMENT
Its function is to sustain the anther.

STYLE
Some styles are solid, others hollow. Their number depends on the number of carpels. The pollen tube grows through the style. In corn the tube can reach a length of 15 inches (40 cm).

OVARY
The ovary is found in the receptacle in the base of the gynoecium, inside the carpels. The pollen tube, which conducts the pollen to the ovule, extends to the ovary.

ROOT
In monocotyledons all the roots branch from the same point, forming a kind of dense hair. They are generally superficial and short-lived.

Whorls

Most flowers have four whorls. In a typical flower the outermost whorl is the calyx, followed by the corolla, the androecium (which can have two parts), and the gynoecium. When a flower has all four whorls, it is considered complete; it is incomplete when it lacks at least one of them. Plants that have an androecium and a gynoecium, but in separate flowers, are called monoecious. If the flower lacks a sepal and petals, it is said to be naked.

250,000

The number of known species of angiosperm plants, thought only 1,000 species have economic importance. About two thirds of these species are native to the tropics.

COROLLA
A grouping of petals. If its parts are separated, they are simply called petals; if they are united, the plant is described as gamopetalous.

PETAL
It typically has a showy color to attract pollinating insects or other animals.

CALYX

The grouping of sepals that protects the other parts of the flower. Together with the corolla it forms the perianth. The sepals may be separate or united; in the latter case the plant is called gamosepalous.

SEPAL
Each of the modified leaves that protect the flower in its first stage of development. They also prevent insects from gaining access to the nectar without completing their pollinating function. Sepals are usually green.

TEPAL
In monocotyledonous plants the petals and sepals are usually the same. In this case they are called tepals, and the group of tepals is called a perianth.

Pollination

The orchid, whose scientific name *Ophrys apifera* means "bee orchid," is so called because of the
similarity between the texture of its flowers and the body of a bee. Orchids' flowers are large
and very colorful, and they secrete a sugary nectar that is eaten by many insects. The
orchid is an example of a zoomophilous species; this means that its survival is based
on attracting birds or insects that will transport its pollen to distant flowers
and fertilize them.

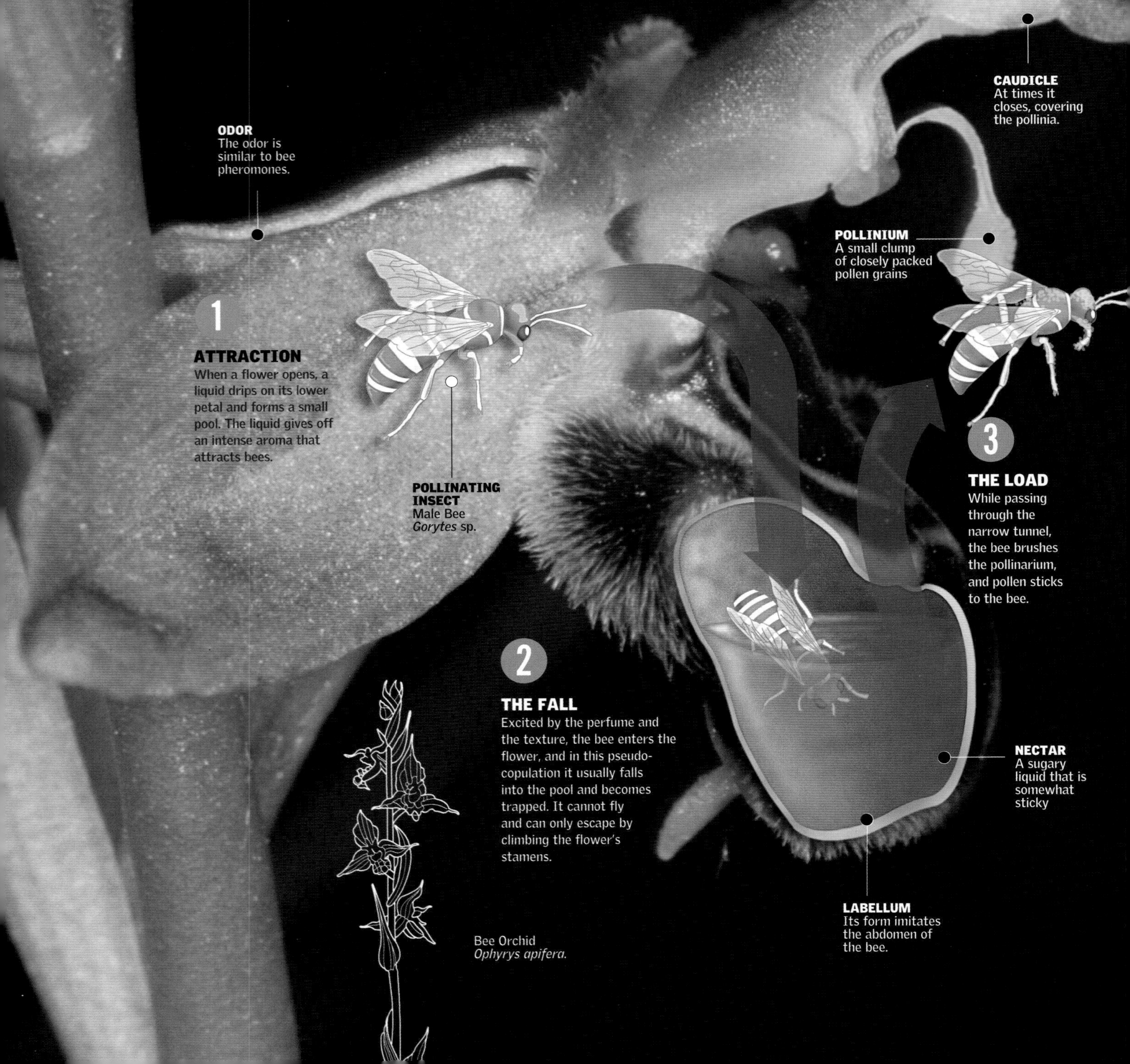

ODOR
The odor is
similar to bee
pheromones.

CAUDICLE
At times it
closes, covering
the pollinia.

POLLINIUM
A small clump
of closely packed
pollen grains

1

ATTRACTION
When a flower opens, a
liquid drips on its lower
petal and forms a small
pool. The liquid gives off
an intense aroma that
attracts bees.

**POLLINATING
INSECT**
Male Bee
Gorytes sp.

3

THE LOAD
While passing
through the
narrow tunnel,
the bee brushes
the pollinarium,
and pollen sticks
to the bee.

2

THE FALL
Excited by the perfume and
the texture, the bee enters the
flower, and in this pseudo-
copulation it usually falls
into the pool and becomes
trapped. It cannot fly
and can only escape by
climbing the flower's
stamens.

NECTAR
A sugary
liquid that is
somewhat
sticky

LABELLUM
Its form imitates
the abdomen of
the bee.

Bee Orchid
Ophyrys apifera.

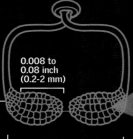

POLLINIA
Small clumps of pollen grains housed in a compartment of the anther.

0.008 to 0.08 inch (0.2-2 mm)

POLLINARIUM
Grouping of two, four, six, or eight pollinia.

GRAIN OF POLLEN

Pollen
Each grain contains a male gamete.

12,000
The number of seeds that a single fertilized orchid produces.

CORBICULUM
Organ for the transport of pollen

COLORATION
is one of the factors of attraction.

4

TRANSFER
The bee takes off toward other flowers, with pollen from the orchid stuck to its back.

5

TOWARD A DESTINATION
When it arrives at another flower of the same species, the bee repeats the incursion and bumps the flower's stigmas (female organs), depositing pollen that is capable of fertilizing it.

LOBULES
They have fine, silky hairs that attract the bees.

CAMOUFLAGE
Some plants that rely on insects for pollination acquire the appearance of the animal species on which they depend for survival. Each orchid has its own pollinating insect.

Bearing Fruit

Once the flower is fertilized, its ovary matures and develops, first to protect the seed forming within it and then to disperse the seed. The stigmas and anthers wither, and the ovary transforms into fruit. Its wall forms the cover, or pericarp. Fruits and seeds are of great economic importance because of their key role in human nutrition. The endosperms of some seeds are rich in starch, proteins, fats, and oils.

Simple Fruits

Come from a single flower. They may contain one or more seeds and be dry or fleshy. Among them are drupes, berries, and pomes.

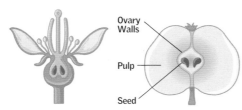

Ovary Walls

Pulp

Seed

Ⓐ POMES
are fleshy fruits that come from epigynous flowers, or flowers whose enclosed ovaries lie below the place where the other parts of the flower are attached. The floral receptacle thickens and forms an edible mesicarp. Apples are one example.

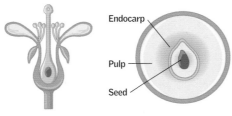

Endocarp

Pulp

Seed

Ⓑ DRUPES
are fleshy fruits, leathery or fibrous, which are surrounded by a woody endocarp with a seed in its interior. They are generally derived from hypogynous flowers—flowers whose ovaries lie above the point where the other flower parts are attached. An example is the peach.

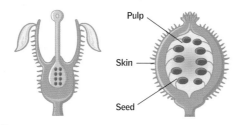

Pulp

Skin

Seed

Ⓒ BERRIES
When they mature, berries generally have a bright color and a fleshy or juicy mesocarp. They come from either epigynous or hypogynous flowers. The grape is an example.

ORANGES
Like other citrus fruits, oranges are similar to berries. Their seeds may propagate when the fruit rots and exposes them or when an animal eats the fruit and then defecates the seeds.

14%
The proportion of an immature citrus fruit that is made up of the Flavonoid Glycoside (Hesperidin).

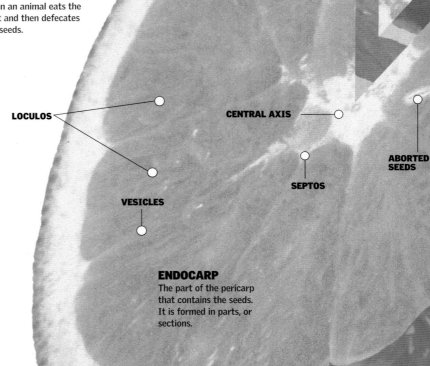

SEEDS

LOCULOS

CENTRAL AXIS

ABORTED SEEDS

SEPTOS

VESICLES

ENDOCARP
The part of the pericarp that contains the seeds. It is formed in parts, or sections.

MESOCARP
A fleshy structure that is relatively solid.

SECTION
A sac that fills with juices
(reserves of water and sugar)
produced by the ovary walls.

PEEL
It consists of the mesocarp and
exocarp of the fruit. It is soft and
secretes oils and acids. However, in
the case of a nut, its hard "peel" is
its endocarp.

MULTIPLE FRUITS

are those that develop from the carpels of more than one
flower, in a condensed inflorescence. When they mature,
they are fleshy. An example is the fig.

Fig
Condensed fruit

Blackberry
In this aggregate fruit,
each berry is a fruit.

Ⓐ AGGREGATE FRUIT
The fruit is made of
numerous drupelets that
grow together.

Ⓑ SYCONIUM
The fruit axis dilates
and forms a concave
receptacle with the shape
of a cup or bottle.

Dry Fruits

are simple fruits whose pericarps dry as they
mature. They include follicles (magnolias),
legumes (peanuts, fava beans, peas), pods
(radishes), and the fruits of many other
species, including the majority of cereals and
the fruits of trees such as maple and ash. Most
dehiscent fruits (fruits that break open to
expose their seeds) are dry fruits.

Mesocarp   Exocarp

Endocarp

EXOCARP
The skin, or external
part, of the fruit.

MOSSES
These plants grow in
wetlands and are devoid of
vascular tissues (roots, stems
and true leaves).

CHAPTER 7

THE OTHER PLANTS

Algae, fungi, lichens, mosses and ferns are different types of plants classified within the group of Talofilas, Bryophytes and Teridofilas, respectively. Earth's most simple plants, they are characterized by reproduction via spores.

Colors of Life

Algae are living things that manufacture their own food using photosynthesis. Their color is related to this process, and it has been used as a way of classifying them. They are also grouped according to the number of cells they have. There are many kinds of one-celled algae. Some algae form colonies, and others have multicellular bodies. Some types of brown seaweed can reach a length of more than 150 feet (45 m).

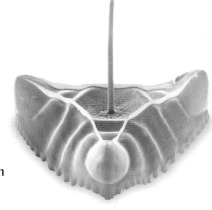

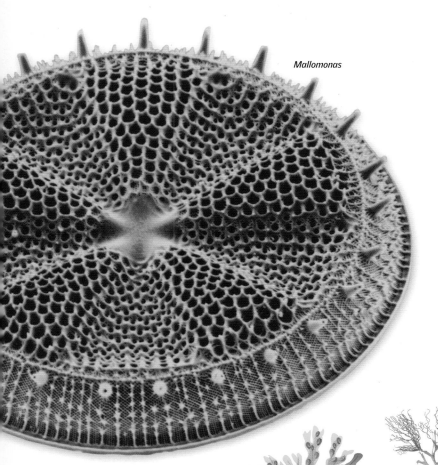

Mallomonas

Single-Celled Organisms

often have flagella that enable them to move through the water. Most have the ability to ingest solid material through phagocytosis. Single-celled algae include some distinctive groups. Diatoms are covered with a protective shell made of silicon. Some single-celled algae, namely red algae, can thrive at relatively high temperatures. Red algae is unique among eukaryote organisms in its ability to live inside thermal water vents.

GREAT OPPORTUNISTS

Single-celled algae live near the surface of bodies of water. When they find an area with light and the nutrients necessary for development, they use asexual reproduction to multiply and colonize the area.

 1

PHAEOPHYTES
are the 1,500 species of brown seaweed. They inhabit temperate regions and the rocky coasts of the coldest seas on Earth. Their color comes from the pigment fucoxanthin, a xanthophyll that masks the green color of their chlorophyll.

Fucus vesiculosus

Dictyota dichotoma implexa

Dictyota dichotoma hudson lamouroux

Cystoseira amantacea stricta

Ectocarpus siliculosus

Multicelled Organisms

This group of algae includes multicelled structures. They form colonies with mobile, single-celled algae that group together more or less regularly in a shared mucilaginous capsule. They can also appear in threadlike shapes, which branch off, or in bulky shapes, which are made up of layers of cells with a particular degree of cellular differentiation, that together are called a thallus.

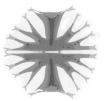

Micrasteria rotata

Scenedesmus quadricauda

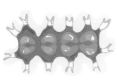

Micrasteria staurastrum

Acetabularia crenulata

Pinnularia borealis

2 **CHLOROPHYTES**

constitute the group of green algae. The majority of species are microscopic, single-celled organisms with flagella. Others form into filaments, and yet others form large multicellular bodies. The group Ulvophyceae includes sea lettuce, which resembles a leaf of lettuce and is edible. The group Charophyceae includes stoneworts, which contain calcium carbonate deposits. The chlorophytes are linked evolutionally with plants because they contain the same forms of chlorophyll, and their cell walls contain cellulose.

Chlamydomonas

6,000

different species have been classified within this group of green algae, or chlorophytes.

3 **RHODOPHYTES**

are characterized by their phycoerythrin pigments, which give the algae a reddish color by masking their chlorophyll's green color. Most rhodophytes grow below the intertidal zone near tropical and subtropical coasts. They are distributed throughout the principal oceans of the world and grow mainly in shaded areas in warm, calm water.

Carrageen red seaweed

Hypoglossum hypoglossoides

Bangia atropurpurea

Nitophyllum punctatum

Halymenia floresia

Apoglossum ruscifolium

How Algae Reproduce

The reproduction of algae can be sexual or asexual in alternating phases, depending on the species and on environmental conditions. Vegetative multiplication occurs through fragmentation or through the production of spores. In sexual reproduction the fertilization of the gametes (sexual cells) produces a zygote that will give rise to a new alga. During asexual reproduction there is no genetic exchange, and the algae produced are clones of the original. Sexual reproduction, in contrast, produces algae with new characteristics that may help them to better adapt to their environment.

Asexual

Asexual reproduction does not involve fertilization. It can take place in either of two ways. In fragmentation, segments of an alga become detached from its body, and, since the alga does not have any specialized organs, the segments continue to grow as long as environmental conditions remain favorable. The other form of asexual reproduction is by means of spores, special cells that form from a normal cell. Some algae spores have one or more filaments, or flagella, that allow the alga to swim freely. When the appropriate environmental conditions are found, the spores germinate into new algae.

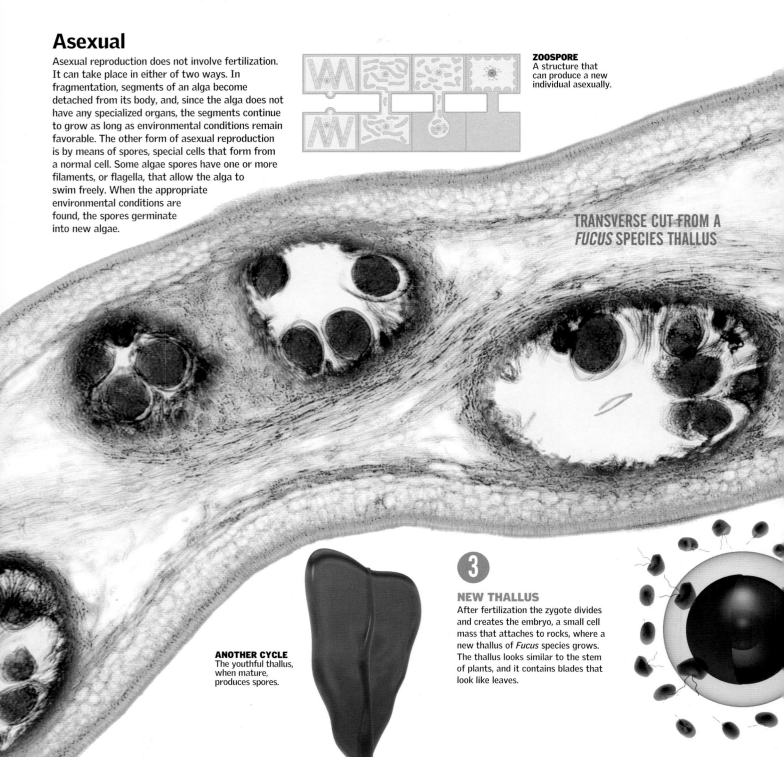

ZOOSPORE
A structure that can produce a new individual asexually.

TRANSVERSE CUT FROM A *FUCUS* SPECIES THALLUS

ANOTHER CYCLE
The youthful thallus, when mature, produces spores.

3

NEW THALLUS
After fertilization the zygote divides and creates the embryo, a small cell mass that attaches to rocks, where a new thallus of *Fucus* species grows. The thallus looks similar to the stem of plants, and it contains blades that look like leaves.

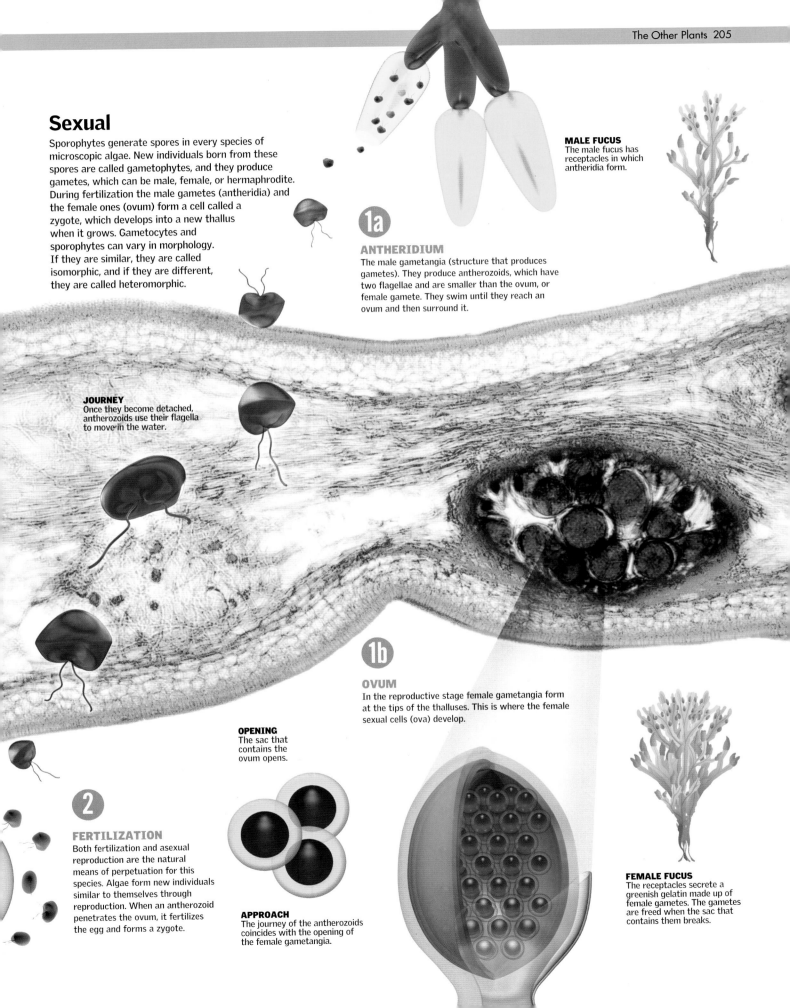

Sexual

Sporophytes generate spores in every species of microscopic algae. New individuals born from these spores are called gametophytes, and they produce gametes, which can be male, female, or hermaphrodite. During fertilization the male gametes (antheridia) and the female ones (ovum) form a cell called a zygote, which develops into a new thallus when it grows. Gametocytes and sporophytes can vary in morphology. If they are similar, they are called isomorphic, and if they are different, they are called heteromorphic.

MALE FUCUS
The male fucus has receptacles in which antheridia form.

1a ANTHERIDIUM

The male gametangia (structure that produces gametes). They produce antherozoids, which have two flagellae and are smaller than the ovum, or female gamete. They swim until they reach an ovum and then surround it.

JOURNEY
Once they become detached, antherozoids use their flagella to move in the water.

1b OVUM

In the reproductive stage female gametangia form at the tips of the thalluses. This is where the female sexual cells (ova) develop.

OPENING
The sac that contains the ovum opens.

2 FERTILIZATION

Both fertilization and asexual reproduction are the natural means of perpetuation for this species. Algae form new individuals similar to themselves through reproduction. When an antherozoid penetrates the ovum, it fertilizes the egg and forms a zygote.

APPROACH
The journey of the antherozoids coincides with the opening of the female gametangia.

FEMALE FUCUS
The receptacles secrete a greenish gelatin made up of female gametes. The gametes are freed when the sac that contains them breaks.

Terrestrial and Marine Algae

As long as there is water, the survival of an alga is assured. Algae are found both in the oceans and in freshwater, but not all can survive in both environments. Depth, temperature, and salt concentrations of water are characteristics that determine whether algae can live in a given area. Algae can be green, brown, or red. Of the three, red algae are found in the deepest waters. Some species of algae can live outside of water, but they are nevertheless found in humid places, such as in mud or on stone walls or rocks.

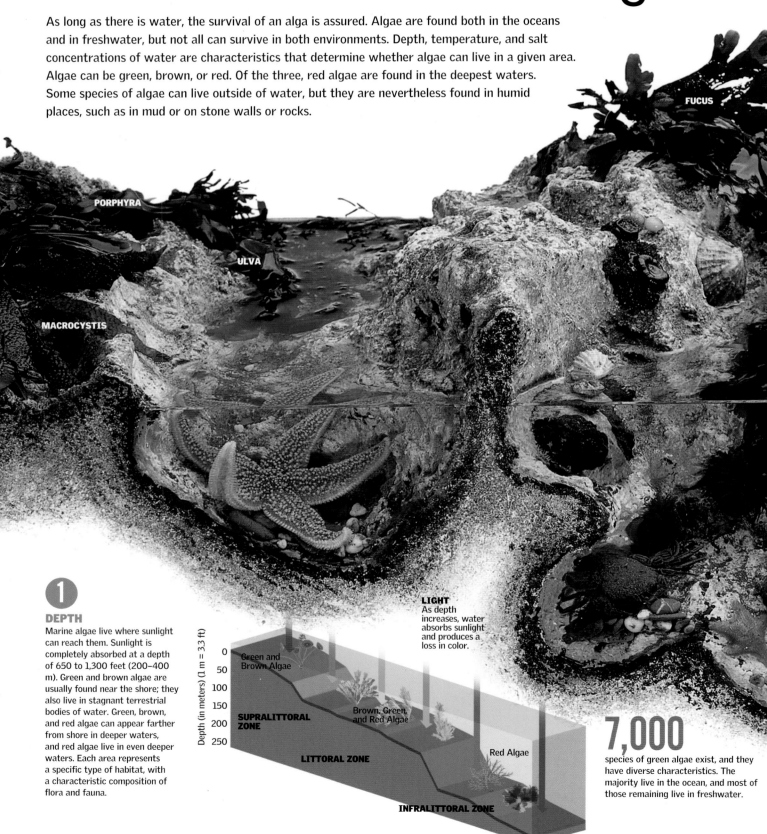

FUCUS

PORPHYRA

ULVA

MACROCYSTIS

1 DEPTH

Marine algae live where sunlight can reach them. Sunlight is completely absorbed at a depth of 650 to 1,300 feet (200–400 m). Green and brown algae are usually found near the shore; they also live in stagnant terrestrial bodies of water. Green, brown, and red algae can appear farther from shore in deeper waters, and red algae live in even deeper waters. Each area represents a specific type of habitat, with a characteristic composition of flora and fauna.

LIGHT
As depth increases, water absorbs sunlight and produces a loss in color.

Depth (in meters) (1 m = 3.3 ft)

0
50
100
150
200
250

Green and Brown Algae

Brown, Green, and Red Algae

Red Algae

SUPRALITTORAL ZONE

LITTORAL ZONE

INFRALITTORAL ZONE

7,000

species of green algae exist, and they have diverse characteristics. The majority live in the ocean, and most of those remaining live in freshwater.

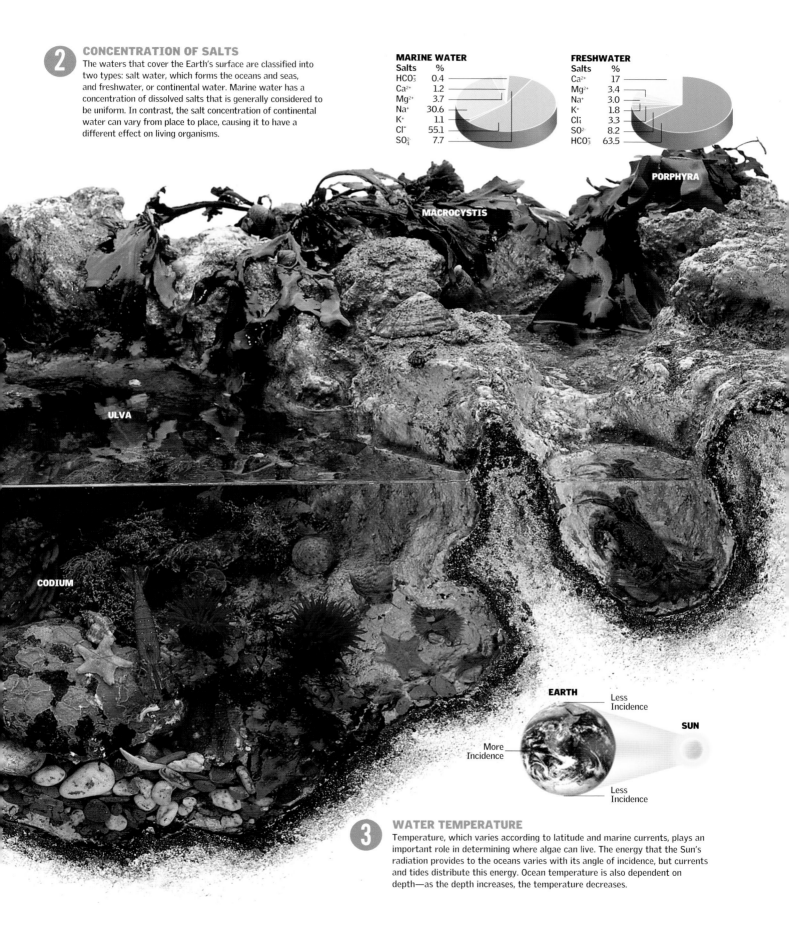

❷ CONCENTRATION OF SALTS

The waters that cover the Earth's surface are classified into two types: salt water, which forms the oceans and seas, and freshwater, or continental water. Marine water has a concentration of dissolved salts that is generally considered to be uniform. In contrast, the salt concentration of continental water can vary from place to place, causing it to have a different effect on living organisms.

MARINE WATER

Salts	%
HCO_3^-	0.4
Ca^{2+}	1.2
Mg^{2+}	3.7
Na^+	30.6
K^+	1.1
Cl^-	55.1
SO_4^{2-}	7.7

FRESHWATER

Salts	%
Ca^{2+}	17
Mg^{2+}	3.4
Na^+	3.0
K^+	1.8
Cl_4^-	3.3
SO^{2-}	8.2
HCO_3^-	63.5

PORPHYRA

MACROCYSTIS

ULVA

CODIUM

EARTH

Less Incidence

SUN

More Incidence

Less Incidence

❸ WATER TEMPERATURE

Temperature, which varies according to latitude and marine currents, plays an important role in determining where algae can live. The energy that the Sun's radiation provides to the oceans varies with its angle of incidence, but currents and tides distribute this energy. Ocean temperature is also dependent on depth—as the depth increases, the temperature decreases.

Strange Bedfellows

Lichens are the result of a close relationship between fungi and algae (usually green algae). Although they are most common in cold areas, they adapt easily to diverse climatic conditions. Lichens can grow in the Arctic glacial regions, as well as in deserts and volcanic regions. They live on rocks, from which they obtain all the necessary minerals to live, and they contribute to the formation of soils. Lichens are excellent indicators of the level of environmental pollution, since elevated levels of pollution cause them to die.

FRUCTICOSE
The long-branched thallus is raised or hanging and can resemble small trees or entangled bushes.

Pseudevernia sp.

0.08 to
0.15 inch
(2-4 mm)

IN THE MOUNTAINS
this lichen is common on the bark of mountain conifers. Its thallus looks like horns.

STIPES
The stipes are projections on the surface of the thallus at which vegetative multiplication takes place. Their shape is variable, and their color may be the same as or slightly darker than that of the thallus.

15,000
classes of lichens exist.

8 inch (2 cm)
The amount a lichen can grow in a year.

4,000 years
The life span a lichen can achieve.

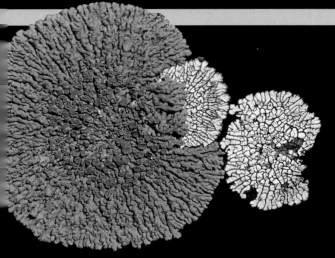

LIKE SCALES

With an appearance of scales, tightly affixed to the substratum, they can be continuous or fragmented in plates or areolas.

Physcia caesia

0.04 to 0.08 inch (1-2 mm)

Where They Live

Lichens grow in cold regions, as well as in the Amazon Rainforest and the desert. They are very sensitive to environmental pollution.

Corticolas
In trunks and branches.

Terricolas
In the soil of forests.

Saxicola
On rocks and walls.

A Symbiotic Relationship

Lichens are the result of symbiosis between a fungus and an alga, a relationship from which both benefit. In a lichen the fungus offers the alga support and moisture and protects it from heat and dehydration. Likewise, the alga produces food for itself and for the fungus through photosynthesis.

HOW IT IS CREATED

1 The spore of the fungus encounters the alga.

Hypha

Germinating Spore

Alga Cell

2 The spore grows around the alga, and the alga reproduces.

3 They form a new organism (thallus of the lichen).

FOLIACEOUS

A showy lichen that has the appearance of widely spread leaves. It is the most common macrolichen.

Lobaria pulmonaria

0.1 to 0.2 inch (3-6 mm)

APOTHECIA
Intervenes in the reproduction of the fungus because it contains its spores.

SOREDIA
Unit of lichen dispersion, formed by groups of gonidia surrounded by hyphae.

HAIRS
Formed by the ends of the hyphae of the cortex or medulla.

LAYER OF ALGAE
The layer contains green algae, which carry out photosynthesis to feed the fungus.

GONIDIA
Name given to algae when they form part of a lichen.

LAYER OF FUNGI
The fungi are generally ascomycetes. They provide the alga with the moisture it needs to live.

HYPHAE
Fungal filaments, which are interwoven and colorless.

MEDULLA
Made up of fungus hyphae.

RICIN
Fixation organs that arise from the cortex or from the medulla.

CORTEX
External layer of the lichen.

Mosses

Mosses were among the earliest plants to emerge. They evolved from green algae more than 250 million years ago and belong to the group of simple plants called bryophytes. Mosses reproduce only in environments where liquid water is present. Because they grow in groups, they take on the appearance of a green carpet. These primitive plants can serve as indicators of air pollution, and they help reduce environmental degradation.

Fertilization

Reproductive organs that produce gametes develop in the green gametophytes, which live all year long. When there is sufficient moisture, the male gamete reaches a female gamete and fertilizes it. The zygote that arises from this union grows and forms the sporophyte. The sporophyte possesses fertile tissue that undergoes meiosis to generate spores that, after falling to the ground and germinating, will form a new gametophyte.

ZYGOTE
It forms from the union of two sexual cells in a watery environment.

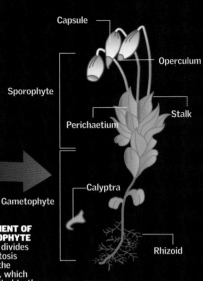

Capsule

Operculum

Sporophyte

Perichaetium

Stalk

Calyptra

Gametophyte

Rhizoid

DEVELOPMENT OF THE SPOROPHYTE
The zygote divides through mitosis and forms the sporophyte, which remains united to the gametophyte.

ADULT SPOROPHYTE
The adult sporophyte consists of a capsule (within which the spores form), a stalk (which holds the capsule), and a foot.

DIPLOID

Diploid cells have two sets of chromosomes. Consequently, they have duplicate genetic information.

Archegonium: the female sexual organ.

Ovule

Antheridium: the male sexual organ.

Spermatozoids

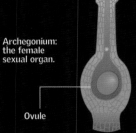

ADULT GAMETOPHYTE
This is what a grown gametophyte looks like.

HAPLOID

A haploid cell is one that contains only one complete set of genetic information. Reproductive cells, such as the ova and sperm in mammals, are haploid, but the rest of the cells in the body of higher organisms are usually diploid—that is, they have two complete sets of chromosomes. In fertilization two haploid gametes unite to form a diploid cell. In the case of mosses all the cells of the gametophyte, the gametes, and the spores are haploid.

The Cycle of Life

Mosses do not have flowers, seeds, or fruits. As with other plants, mosses have a life cycle formed by alternating generations; however, in contrast with vascular plants, the haploid gametophyte is larger than the diploid sporophyte. Their biological cycle begins with the release of spores, which form in a capsule that opens when a small cap called the operculum is ejected. The spores germinate and give rise to a filamentous protonema (cellular mass) from which the gametophyte develops. The zygote that forms from the union of the two sexual cells develops into the sporophyte.

GAMETOPHYTE DEVELOPMENT
The gametophyte grows.

HORIZONTAL FILAMENTS
The gametophyte develops from the horizontal filaments.

GERMINATION OF THE SPORE
The spore germinates and gives rise to a filamentous protonema (cellular mass).

Rhizoids

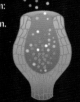

ANNULUS

OPERCULUM
A type of cap that covers the opening of the capsule and normally separates when the spores exit.

Meiosis

Meiosis is a type of cellular division in which each daughter cell receives only one complete set of chromosomes. Therefore, the resulting cells have half as many chromosomes as the parent cells had. In general, this mechanism generates the gametes, but mosses generate haploid spores in the capsule of the sporophyte.

Mature Sporophyte
consists of a capsule in which spores are formed.

SPORES
The life cycle of a moss begins with the freeing of the spores that form in the capsule, which opens when a cap called the operculum is expulsed.

FUNARIA HIGROMETRICA
belongs to the group of plants called bryophytes.

10,000
species of mosses have been classified within the bryophite group of nonvascular plants.

CAPSULE
contains the spores and is found at the tip.

Small Plants

Mosses are bryophytes. They are relatively small plants that affix themselves to a substratum via rhizoids and carry out photosynthesis in small "leaves" that lack the specialized tissues of the real leaves of vascular plants. They fulfill a very important ecological role: they participate in the formation of soils by decomposing the rocks on which they grow, and they contribute to the photosynthesis of epiphytes in rainforests. Their asexual reproduction occurs through fragmentation or the production of propagula.

SPOROPHYTE
The sporophyte does not have an independent existence but lives at the expense of the gametophyte. The sporophyte lives a short time and only during a certain time of the year.

0.2 inch
(5 mm)

Dispersion of Spores

The fern is one of the oldest plants. Ferns have inhabited the surface of the Earth for 400 million years. Their leaves have structures called sori that contain the sporangium, which houses the spores. When the sori dry up, they release the spores into the air. Once on the ground, the spores germinate as gametophytes. In times of rain and abundant moisture the male cells of the gametophyte are able to swim to reach female gametes, which they fertilize to form a zygote that will grow as a sporophyte.

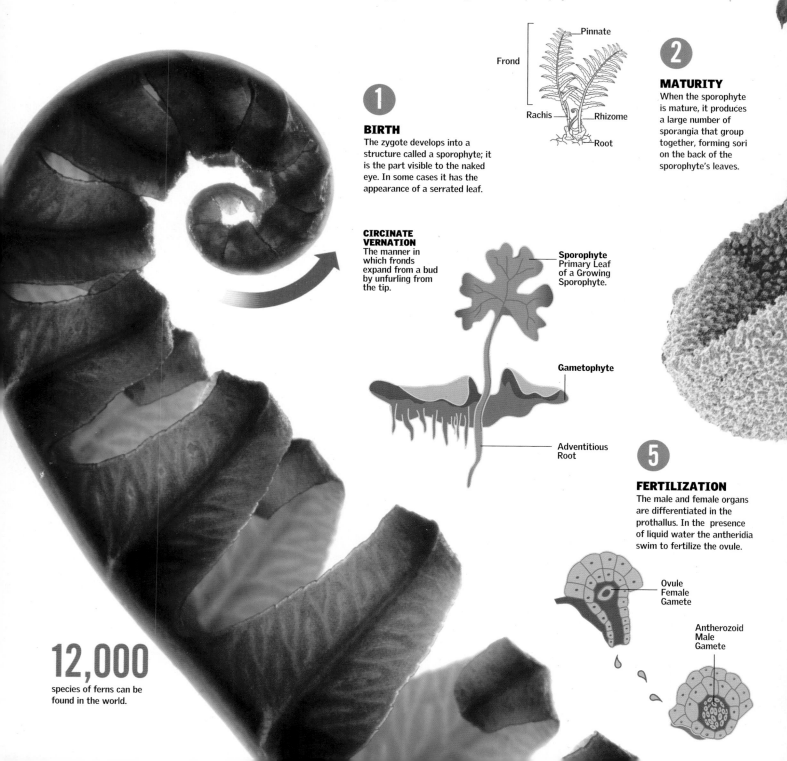

1

Frond — Pinnate

Rachis — Rhizome

— Root

BIRTH
The zygote develops into a structure called a sporophyte; it is the part visible to the naked eye. In some cases it has the appearance of a serrated leaf.

CIRCINATE VERNATION
The manner in which fronds expand from a bud by unfurling from the tip.

2

MATURITY
When the sporophyte is mature, it produces a large number of sporangia that group together, forming sori on the back of the sporophyte's leaves.

Sporophyte
Primary Leaf of a Growing Sporophyte.

Gametophyte

Adventitious Root

5

FERTILIZATION
The male and female organs are differentiated in the prothallus. In the presence of liquid water the antheridia swim to fertilize the ovule.

Ovule Female Gamete

Antherozoid Male Gamete

12,000
species of ferns can be found in the world.

PINNULES
Smaller lobes that contain sori on their inner side.

SORI
Contains the sporangia.

INDUSIUM
Small cap that protects and covers the sori while the spores mature inside each sporangium.

PLACENTA

SPORANGIUM
Microscopic capsule that contains the spores.

FILAMENT
unites with the pinnule in the placenta.

PINNAS
Petioles into which the leaf is divided.

3

CATAPULT OF SPORES
When the sporangia dry and wither, they liberate spores through a catapult mechanism.

300 million
The number of spores one fern lead can produce. Their total weight is 0.04 ounce (1 G).

SPORE
The spore is the most effective unit of dispersion because of its aerodynamic form and microscopic dimensions.

THIN WALL
Formed by a single layer of cells.

ANNULUS
Row of cells located on the back wall. When it dries, the number of sporangia doubles.

4

GERMINATION
When the spore encounters the right environment, it develops into a multicellular structure that forms the haploid gametophyte, called the prothallus.

Atheridium
Male Sex
Organ

Archegonium
Female Sex
Organ

GAMETOPHYTE

Rhizoid

YOUNG PROTHALLUS

Cellular sheet that forms the prothallus

Rhizoid

Spore

HOW A LEPTOSPORANGIUM IS FORMED

A
It starts as a single initial epidermal cell.

B
The lower cell gives rise to a thin stalk.

C
The stalk divides into four initial cells and small sporocytes.

D
The wall of the mature sporangium is formed by a single layer of cells.

E
It forms a fixed number of spores through meiosis.

Another World

For many years fungi were classified within the plant kingdom. However, unlike plants, they are heterotrophic—unable to produce their own food. Some fungi live independently, whereas others are parasitic. Like animals, they use glycogen for storing reserves of energy, and their cell walls are made of chitin, the substance from which insects' outer shells are made.

Fungi: A Peculiar Kingdom

Fungi can develop in all sorts of environments, especially damp and poorly lit places, up to elevations of 13,000 feet (4,000 m). They are divided into four large phyla, in addition to a group of fungi called "imperfect" because they generally do not reproduce sexually. At present, 15,000 species of fungi fall into this category. DNA analysis has recently reclassified them as Deuteromycetes.

CHYTRIDIOMYCOTA

are the only fungi that, at some point in their lives, have mobile cells—male and female gametes, which they release into water in order to reproduce. They live in water or on land, feeding on dead material or living as parasites on other living organisms. Their cell walls are made of chitin.

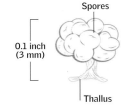

Spores

0.1 inch
(3 mm)

Thallus

VARIETY
There are great anatomical differences among the Chytridiomycetes. In the same reproductive phase they can produce haploid and diploid spores.

39° to
140° F
(4°-60° C)

The temperature range in which most fungi can live in humid climates.

SPORES

SLIME MOLD
Physarum polycephalum.

DEUTEROMYCOTA

Are also called "imperfect fungi" because they are not known to have a form of sexual reproduction. Many live as parasites on plants, animals, or humans, causing ringworm or mycosis on the skin. Others—such as *Penicillium*, which produces penicillin, and *Cyclospora*—have great medicinal and commercial value.

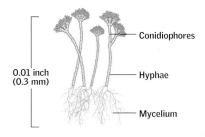

0.01 inch
(0.3 mm)

Conidiophores

Hyphae

Mycelium

OF UNKNOWN SEX
In Deuteromycetes, conidia are tiny spores that function asexually. They are contained in structures called conidiophores.

BASIDIOMYCOTA

This phylum, which includes mushrooms, is the most familiar of the fungi. The mushroom's reproductive organ is its cap. Its branches grow underground or into some other organic substrate.

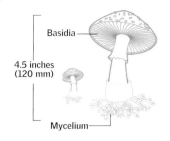

Basidia

4.5 inches
(120 mm)

Mycelium

CAPPED MUSHROOMS

With its recognizable shape, the mushroom's cap protects the basidia, which produce spores.

CHANTERELLE MUSHROOM
Cantharellus cibarius.

BLACK BREAD MOLD
Rhizopus nigricans.

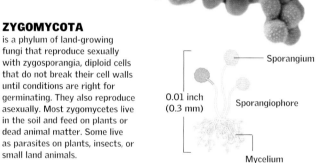

80,857

different species have been identified in the fungi kingdom. There are believed to be approximately 1,500,000 species.

ZYGOMYCOTA

is a phylum of land-growing fungi that reproduce sexually with zygosporangia, diploid cells that do not break their cell walls until conditions are right for germinating. They also reproduce asexually. Most zygomycetes live in the soil and feed on plants or dead animal matter. Some live as parasites on plants, insects, or small land animals.

Sporangium

0.01 inch
(0.3 mm)

Sporangiophore

Mycelium

MANY LITTLE POUCHES

Its spores are formed when two gametes of opposite sexes fuse. It can also reproduce asexually, when the sporangium breaks and releases spores.

FRUITING BODIES

WHITE MYCELIUM

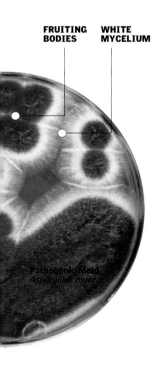

Pathogenic Mold
Aspergillus niger

ASCUS WITH ASCOSPORES

ASCOMYCOTA

is the phylum with the most species in the fungi kingdom. It includes yeasts and powdery mildews, many common black and yellow-green molds, morels, and truffles. Its hyphae are partitioned into sections. Their asexual spores (conidia) are very small and are formed at the ends of special hyphae.

ERGOT
Claviceps purpurea.

EXPLOSIVE

At maturity the asci burst. The explosion releases their sexual spores (ascospores) into the air.

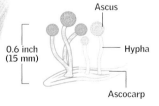

Ascus

0.6 inch
(15 mm)

Hypha

Ascocarp

The Diet of Fungi

Fungi do not ingest their food like animals. On the contrary, they absorb it after breaking it down into small molecules. Most of them feed on dead organic material. Other fungi are parasites, which feed on living hosts, or predators, which live off the prey they trap. Many others establish relationships of mutual benefit with algae, bacteria, or plants and receive organic compounds from them.

CAP
Besides being easy to spot, the cap is the fertile part of basidiomycetes; it contains spores.

Chemical Transformation

The organic or inorganic substances that fungi feed on are absorbed directly from the environment. Fungi first secrete digestive enzymes onto the food source. This causes a chemical transformation that results in simpler, more easily assimilated compounds. Basidiomycetes are classified according to their diet. For example, they colonize different parts of a tree depending on the nutrients they require.

PARASITES
Fungi such as *Ceratocystis ulmi* and *Agrocybe aegerita* (shaded areas on the leaf) live at the expense of other plants, which they can even kill. Others live parasitically off animals.

SAPROBES
There is no organic material that cannot be broken down by this type of fungus. They actually live on the dead parts of other plants, so they cause no harm to the host.

SYMBIOTIC
While feeding off the plant, they help it to obtain water and mineral salts more easily from the soil. Each species has its own characteristics.

Fungi of the genus *Amanita*, including the poisonous *A. muscaria* shown here, have the well-known mushroom shape with a mushroom cap.

MYCELIUM
When a mushroom spore finds the right medium, it begins to generate a network of hyphae, branching filaments that extend into the surrounding medium. This mass of hyphae is called a mycelium. A mushroom forms when threads of the mycelium are compacted and grow upward to create a fruiting body.

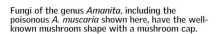

Spore-producing structures

Hyphae

FRUITING BODY
The basidiocarp, or mushroom cap, generates new spores.

VEGETATIVE MYCELIUM
It is made of branches of threadlike hyphae that grow underground.

CUTICLE

The skin, or membrane, that covers the cap, or pileus, is called the cuticle. It can have a variety of colors and textures, such as velvety, hairy, scaly, threadlike, fibrous, fuzzy, smooth, dry, or slimy.

GILLS

are the structures that produce spores. Their shape varies according to the species.

DETAIL OF A GILL

BASIDIA

are fine structures that contain groups of four cells, which are able to reproduce.

Basidium

Basidiospore

HYMENIUM

It is located on the underside of the cap. It contains very fine tissues that produce spores. Its structure can consist of tubes, wrinkles, hairlike projections, or even needles.

LIFE CYCLE OF A FUNGUS

Fungi produce spores during sexual or asexual reproduction. Spores serve to transport the fungus to new places, and some help the fungus to survive adverse conditions.

Development of the fruit-bearing body

Spore formation by fertilization

Hyphae formation

Release of spores

RING

Also known as the veil, it protects part of the hymenium in young fungi.

Growth

At birth the fruiting body of the species *Amanita muscaria* looks like a white egg. It grows and opens slowly as the mushroom's body unfolds. As it grows the cap first appears completely closed. During the next several days it opens like an umbrella and acquires its color.

HALLUCINOGENIC MUSHROOM

Psilocybin aztecorum.

STEM

Cylindrical in shape, it holds up the cap and reveals important information for identifying the species.

Did You Know?

Fungi can break down an impressive variety of substances. For example, a number of species can digest petroleum, and others can digest plastic. Fungi also provided the first known antibiotic, penicillin. They are now a basic source of many useful medical compounds. Scientists are studying the possibility of using petroleum-digesting fungi to clean up oil spills and other chemical disasters.

VOLVA

The volva is made of the remains of the early rings that have fallen off. It differs from species to species.

STROBILURUS ESCULENTUS

lives on the cones of various pine trees.

Poison in the Kingdom

A poisonous fungus is one that, when ingested, causes toxic effects. In terms of its effects on the eater, the toxicity can vary according to the species and to the amount ingested. At times poisoning is not caused by eating fungi but by eating foods, such as cereal products, that have been contaminated by a fungus. Rye, and to a lesser extent oats, barley, and wheat, can host toxic fungi that produce dangerous mycotoxins. These mycotoxins can cause hallucinations, convulsions, and very severe damage in the tissues of human organs.

Attack on Rye

Ergot (*Claviceps purpurea*) is a parasite of rye and produces alkaloid mycotoxins—ergocristine, ergometrine, ergotamine, and ergocryptine. When barley with ergot is processed for use in food, the mycotoxins can be absorbed when eaten. All these toxic substances can act directly on nerve receptors and cause the constriction of blood vessels.

2

FRUIT
The perithecium is a type of fruiting, or reproductive, body in ascomycetes. It is a type of closed ascocarp with a pore at the top. The asci are inside the perithecium.

3

SPORES
The asci are sac-shaped cells that contain spores called ascospores. In general, they grow in groups of eight and are light enough to be scattered into the air.

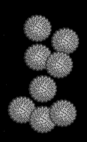

1

RELEASE
Within the enclosing structures a stroma, or compact somatic body, is formed. Inside it reproductive growths develop, which contain a large number of perithecia.

INGESTION
The main means of intake of the mycotoxins is through products manufactured with flour.

Ergotism

Ergotism, or St. Anthony's Fire, is a condition caused by eating products such as rye bread that have been contaminated with alkaloids produced by *Claviceps purpurea* fungi, or ergot. The alkaloids typically affect the nervous system and reduce blood circulation in the extremities, which produces the burning sensation in the limbs that is one of the condition's notable symptoms.

NERVOUS SYSTEM
Lethargy, drowsiness, and more severe conditions, such as convulsions, hallucinations, and blindness, are symptoms caused by the effects of ergot on the nervous system.

EXTREMITIES
Ergotamine alkaloids cause the constriction of blood vessels, leading to gangrene.

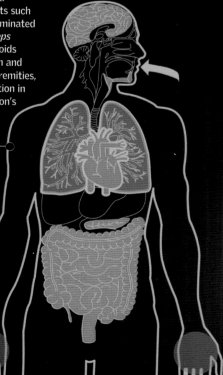

ERGOT
Claviceps purpurea.

4

PARASITES

Ascospores of sexual origin or asexual conidia develop as parasites in the ovary of the rye flower. They cause the death of its tissues and form sclerotia. In some languages ergot's name is related to the word for "horn" because of sclerotia's hornlike shape.

Poisonous Mushrooms

Eating the fruiting bodies of some species can be very dangerous if it is not clearly known which are edible and which are poisonous. There is no sure method for determining the difference. However, it is known for certain that some species—such as certain species of the genera *Amanita*, *Macrolepiota*, and *Boletus*—are poisonous.

DESTROYING ANGEL
Amanita virosa.

PRETTY BUT DEADLY

This mushroom is toxic to the liver. It grows from spring to fall, often in sandy, acidic soil in woodlands and mountainous regions. Its cap is white and 2 to 5 inches (5-12 cm) in diameter. Its stem and gills are also white, and the gills may appear detached from the stem. The base of the stem has a cuplike volva, but it may be buried or otherwise not visible.

INSECTICIDE

The fly agaric's name is thought to come from its natural fly-killing properties. Its cap is typically red and 6 to 8 inches (15-20 cm) in diameter. It may be covered with white or yellow warts, but they are absent in some varieties. The stem is thicker at the base, which looks cottony. It also has a large white ring that looks like a skirt. It grows in summer and fall in coniferous and deciduous forests. If eaten, it causes gastrointestinal and psychotropic symptoms.

FLY AGARIC
Amanita muscaria.

RYE BREAD

WHISKEY

FLOUR

Derived from Rye

In Europe during the Middle Ages wheat bread was a costly food, not part of the common diet. Most people ate bread and drank beer prepared from rye. This made them susceptible to ingesting mycotoxins from *Claviceps purpurea*. Thus, the largest number of cases of ergotism occurred during this time. Today preventative controls in the production of bread and related products from rye and other cereals have greatly reduced instances of ergotism.

Pathogens

Fungi that are able to cause illnesses in people, animals, or plants are called pathogens. The nocive, or toxic, substances that these organisms produce have negative effects on people and cause significant damage to agriculture. One reason these pathogens are so dangerous is their high tolerance to great variations in temperature, humidity, and pH. *Aspergillus* is a genus of fungi whose members create substances that can be highly toxic.

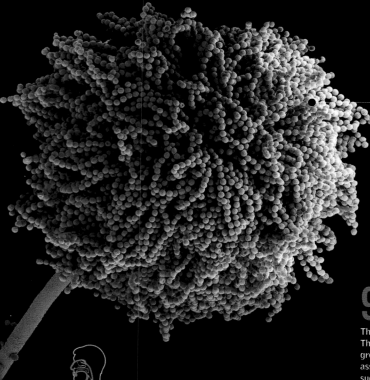

CONIDIA CHAIN
Conidia are asexual spores that form at the ends of the hyphae. In this case they group together in chains.

CONIDIA
are so small that they spread through the air without any difficulty.

PHIALIDES
are cells from which conidia are formed.

900

The number of *Aspergillus* species. They have been classified into 18 groups. Most of these species are associated with human illnesses, such as Aspergillosis.

CONIDIOPHORE
The part of the mycellium of the fruiting, or reproductive, body in which asexual spores, or conidia, are formed

ALLERGENICS
Aspergillus flavus
This species is associated with allergic reactions in people with a genetic predisposition to this allergy. They also cause the contamination of seeds, such as peanuts. They produce secondary metabolites, called mycotoxins, that are very toxic.

SAPROBIA
Aspergillus sp.
In addition to the pathogen species, there are some species of *Aspergillus* that decompose the organic matter of dead insects, thus incorporating nutrients into the soil.

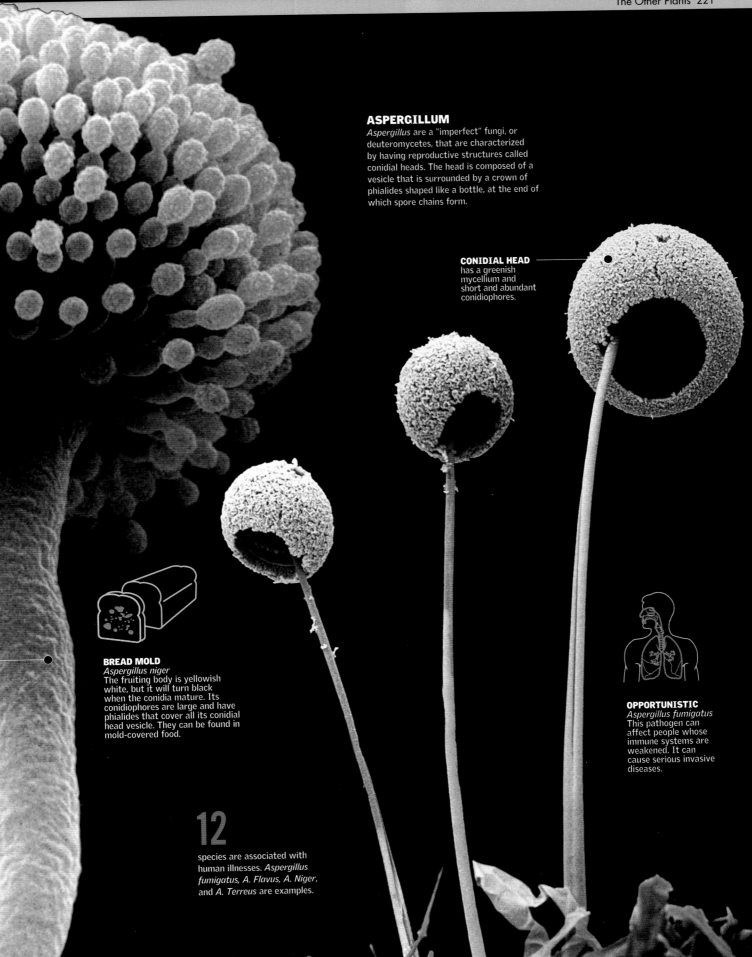

ASPERGILLUM

Aspergillus are a "imperfect" fungi, or deuteromycetes, that are characterized by having reproductive structures called conidial heads. The head is composed of a vesicle that is surrounded by a crown of phialides shaped like a bottle, at the end of which spore chains form.

CONIDIAL HEAD
has a greenish mycellium and short and abundant conidiophores.

BREAD MOLD
Aspergillus niger
The fruiting body is yellowish white, but it will turn black when the conidia mature. Its conidiophores are large and have phialides that cover all its conidial head vesicle. They can be found in mold-covered food.

OPPORTUNISTIC
Aspergillus fumigatus
This pathogen can affect people whose immune systems are weakened. It can cause serious invasive diseases.

12

species are associated with human illnesses. *Aspergillus fumigatus, A. Flavus, A. Niger,* and *A. Terreus* are examples.

Destroying to Build

YEAST
Saccharomyces cerevisiae.

Yeasts, like other fungi, decompose organic material. This capacity can be beneficial, and, in fact, human beings have developed yeast products for home and industrial use, such as bread, baked goods, and alcoholic beverages, that attest to its usefulness. Beer manufacturing can be understood by analyzing how yeasts feed and reproduce and learning what they require in order to be productive.

Precious Gems

Yeast from the genus *Sacchromyces cerevisiae* can reproduce both asexually and sexually. If the concentration of oxygen is adequate, the yeasts will reproduce sexually, but if oxygen levels are drastically reduced, then gemation will take place instead. Gemation is a type of asexual proliferation that produces child cells that split off from the mother cell. Starting with barley grain, this process produces water, ethyl alcohol, and a large quantity of CO_2, the gas that forms the bubbles typically found in beer.

FERMENTATION

Under anaerobic conditions yeasts can obtain energy and produce alcohol. By means of the alcoholic fermentation process they obtain energy from pyruvic acid, a product of the breakdown of glucose by glucolysis. In this process CO_2 is also produced and accumulated, as is ethyl alcohol. The carbon dioxide will be present in the final product: the beer.

WINE YEAST

Yeast is also used to produce wine. In wine production, however, the CO_2 that is produced is eliminated.

2 Spores
A sac called an ascus is formed that contains ascospores of yeast.

1 Meiosis
A diploid cell forms four haploid cells.

Cycle

Grow and Multiply. As long as they have adequate nutrients, yeasts will continuously repeat their reproductive life cycle.

3 Release of the Ascospores
The opening of the ascus releases the spores, which then reproduce by mitosis.

4 Union of the Ascospores
The haploid cells fuse and form a new diploid cell.

5 Gemation
Under the right conditions the diploid cells begin to reproduce asexually.

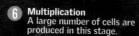

6 Multiplication
A large number of cells are produced in this stage.

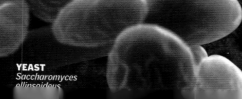

YEAST
Saccharomyces ellipsoideus

Homemade Bread

Many products are made with yeasts, and one of the most important is bread. In the case of bread, yeasts feed off the carbohydrates present in flour. Bread products, unlike alcoholic beverages, need to have oxygen available for the yeast to grow. The fungi release carbon dioxide as they quickly consume the nutrients. The bubbles of carbon dioxide make the dough expand, causing the bread to rise.

NUCLEUS
It coordinates all the cell's activities. Its duplication is vital in making each child cell the same as its progenitor cell.

MITOCHONDRIA
These subcellular structures become very active when the cell is in an environment rich in oxygen.

CELL MEMBRANE
The cell membrane controls what enters or exits the cell. It acts as a selective filter.

GEMATION
Buds, or gems, which will become independent in a new cell, are formed in different parts of a yeast.

12%
The maximum percentage of alcohol that yeast will tolerate.

ENZYME PRODUCTION
Internal membrane systems produce the enzymes that regulate the production of alcohol and carbon dioxide in the cells.

VACUOLE
This organelle contains water and minerals that are used in the cell's metabolism. The concentration of these nutrients helps regulate the activity of the cell.

**GRAND CANYON
NATIONAL PARK**
View of the Grand Canyon
(northern Arizona) from the Yavapai
Point area on the South Rim.

CHAPTER 8

NATURE WONDERS

From the snowy peaks of the Himalayas and Kilimanjaro to the sculpted rocks of the Grand Canyon, from the great waterfalls of Iguazu to the harsh desert climate of the Sahara, there are still some real paradises on our planet, where nature can express herself with all her intensity, and where life emerges in its most varied forms.

The Sahara

The world's largest desert, it covers a vast plain of more than 9,000,000 sq km in northern Africa. This plain is interrupted by rocky plateaus or hamadas, silt and gravel or regs, and seas of sand dunes or ergs. The daytime temperatures that reach 50° C, fall sharply at night, when the record may drop to 30° C below zero.

Tuareg

are people living in the central Sahara, especially in the territories of Mali and Niger. They are characterized by the nomadic way of life, so that the Tuareg families live primarily in nomadic camps.

Little sand

The area covered by sand is only 9% of the total in the Sahara desert. The ergs, extensions of sand moved by the wind, differentiate from the hamada or the stony desert.

Food shortages

Beyond the heat, another major obstacle that the inhabitants of this desert must face is the lack of food. While herbivores must travel long distances to find all the plants they need, the predators, such as the fennec, given the lack of variety of animals, do not easily find their prey, and sometimes can not even afford to choose. The fat-tailed gerbil relies on the occasion when food is abundant to replenish its stores of fat, in such a way that it find is difficult to move.

200 liters

of water is the quantity a camel can reach to drink in one go.

DESERTS IN THE WORLD

A vast belt of desert extends from the Atlantic Ocean to Central Asia, where the Sahara desert stands out. These dry areas are located mainly in tropical and subtropical regions. Here temperatures are very high throughout the year because of high insolation. Consequently the rainfall, albeit extremely rare, can fall only in summer.

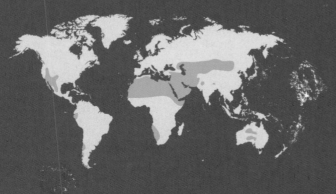

Oasis

The ground water emerges into the ponds where animals drink and date palms grow, in whose shade fruit trees and cereals are grown.

LAPPET-FACED VULTURE
Torgos tracheliotus.

ADDAX
Addax nasomaculatus.

COMMON GUNDI
Ctenodactylus gundi.

DORCAS GAZELL
Gazella dorcas.

ANTLION
Noaleon limbatellus.

MARCHING ON THE SAND
Dromedaries and camels are adapted to very soft sandy soil. Their hooves do not directly support the body weight, which is distributed on its pads, only the front ends of the hoof hit the ground.

DESERT ROSE
This is an unusual sedimentary rock created in the Sahara, which is formed in the shape of a flower.

HAMADA
This type of rocky plateau occupies 80 per cent of the Saharan desert.

NOCTURNAL ACTIVITY
During the day, the oppressive heat forces th e majority of animals to take shelter. The sharp drop in temperature at sundown allows them to go out in search of food and water.

EGYPTIAN PLOVER
Pluvianus aegyptius.

DROMEDARY
Camelus dromedarius. They can drink up to 150 liters of water in one go.

GEMSBOK
Oryx gazella.

FENNEC
Vulpes zerda.

PIN-TAILED SANDGROUSE
Pterocles alchata.

DESERT HEDGEHOG
Hemiechinus auritus.

HELMETED GUINEA FOWL
Numida meleagris.

SPINY-TAILED LIZARD
Uromastyx geyri.

LARGE EGYPTIAN GERBIL
Gerbillus pyramidum.

SCORPION
Androctonus australis.

SANDFISH
Scincus scincus.

BLACK SCARAB
Oxycara gastonis.

MOSSAMBIQUE SPITTING COBRA
Naja mossambica.

Death Valley

In the Mojave Desert, in eastern California, Death Valley spreads over more than 13,000 km² (5,019 sq. miles). The phenomenon known as 'sailing stones' at Racetrack Playa and the 'singing' dunes of Eureka Valley, make this place one of the greatest natural spectacles on Earth.

Attributable to the fault

The 155 mi (250 km) long basin throughout Death Valley is divided in two by a system of right lateral-moving geological faults. The Valley was formed by the process known as "sinking", which explains the fact that it is located dozens of meters beneath sea level.

201º F (94º C)

The highest ground temperature recorded in Death Valley (15 July 1972).

DOWNTHROWN BLOCK

FAULT PLANE

UPTHROWN BLOCK

AIR CIRCULATION

With temperatures reaching around 116° F (47° C) in summer, Death Valley is possibly the hottest place on Earth. The hot air rises over the rugged mountain ranges, where its temperature drops slightly and it sinks into the valley once again, where it heats up once more. This cycle increases the temperature of the soil and further still when currents of hot air are generated; these burn the arid soil and maintain the desert conditions of the valley.

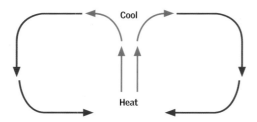

Cool

Heat

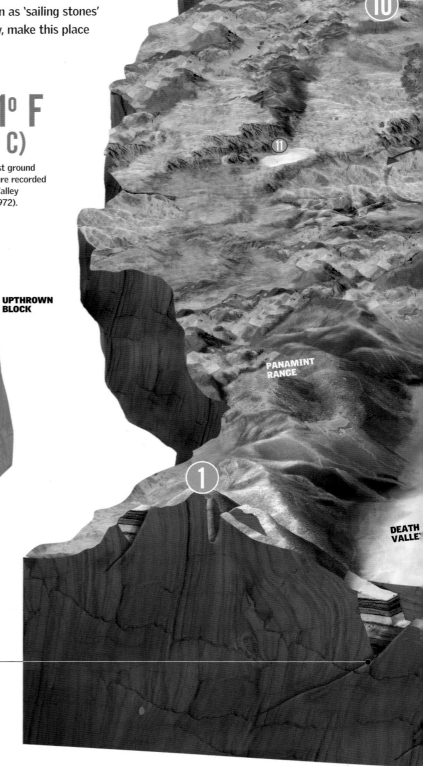

PANAMINT RANGE

DEATH VALLEY

EUREKA DUNES

Measuring 5 km (3.1 miles) long and 1.5 km (0.9 miles) wide, they represent the largest dune field in California. From the summit of the dunes, a unique phenomenon can be witnessed: singing sand. When the sand slides across the slopes, it produces a strange, deep sound, which is reminiscent of a note being played on an organ. These are the most common sand formations:

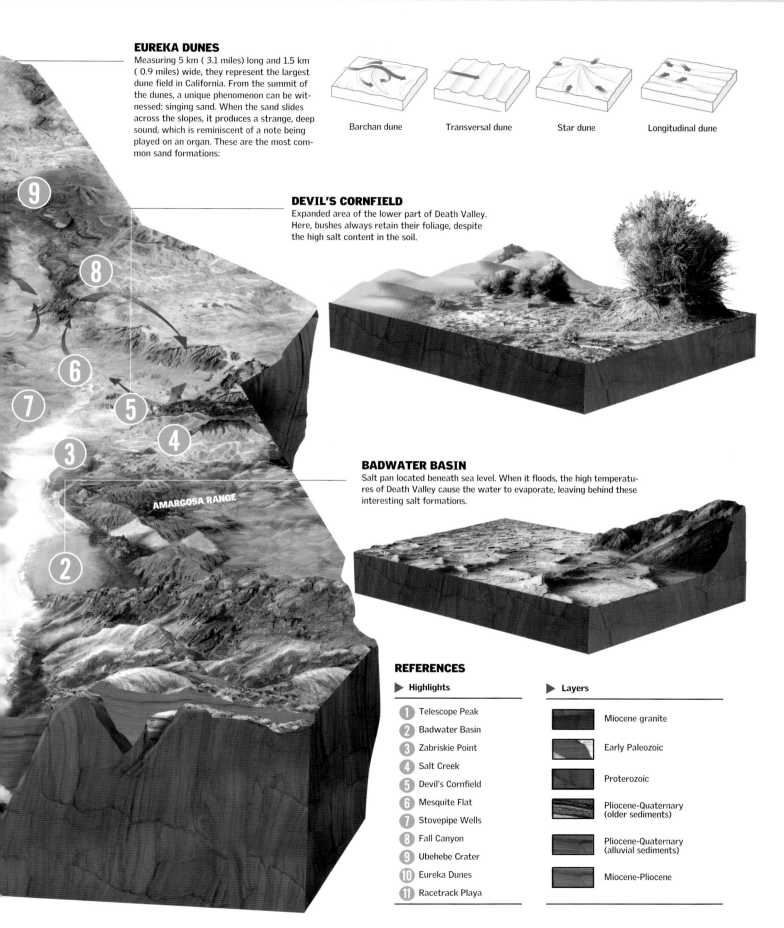

Barchan dune

Transversal dune

Star dune

Longitudinal dune

DEVIL'S CORNFIELD

Expanded area of the lower part of Death Valley. Here, bushes always retain their foliage, despite the high salt content in the soil.

AMARGOSA RANGE

BADWATER BASIN

Salt pan located beneath sea level. When it floods, the high temperatures of Death Valley cause the water to evaporate, leaving behind these interesting salt formations.

REFERENCES

▶ Highlights

1 Telescope Peak
2 Badwater Basin
3 Zabriskie Point
4 Salt Creek
5 Devil's Cornfield
6 Mesquite Flat
7 Stovepipe Wells
8 Fall Canyon
9 Ubehebe Crater
10 Eureka Dunes
11 Racetrack Playa

▶ Layers

Miocene granite

Early Paleozoic

Proterozoic

Pliocene-Quaternary (older sediments)

Pliocene-Quaternary (alluvial sediments)

Miocene-Pliocene

The Himalaya Range

It is justly named 'the root of the world'. The Himalayan chain has the highest mountains on Earth, which rise to almost 9,000 meters. This huge mountain range, about 594,400 sq km, spreads across borders and enlarge across Pakistan, India, Nepal, Bhutan and Tibet, now occupied by China.

The Journey to India

The elevation of the Himalaya is a result of the collision between the Indian and the Eurasian Plates. This clash began about 50 million years ago and deformed the collision zone, raising the Himalayas and forming the Tibetan Plateau as well. As the plate moves, about 2 cm each year northward nowadays, the Himalayas rise more and more.

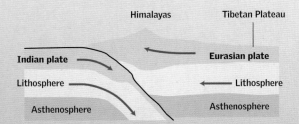

Himalayas · Tibetan Plateau

Indian plate · Eurasian plate

Lithosphere · Lithosphere

Asthenosphere · Asthenosphere

7.87 in
20 cm

The speed at which the India plate had been travelling for about 40 million years before colliding with the Eurasian Plate.

THE GREAT WALL
With peaks exceeding 26,000 ft (8,000 m), the highest mountain range in the world is an almost impassable barrier between the Indian subcontinent and the rest of Asia.

THE HIGHEST MOUNTAINS

8.167 m	8.163 m	8.201 m	8.848 m	8.516 m	8.463 m	8.586 m
DHAULAGIRI (Nepal)	**MANASLU** (Nepal)	**CHO OYU** (China/Nepal)	**EVEREST** (China/Nepal)	**LHOTSE** (China/Nepal)	**MAKALU** (China/Nepal)	**KANCHENJUNGA** (India/Nepal)

STRATA HERBS
Moss, lichens and small shrubs can grow up to 4,500 m (14,750 ft). Up to 3,500 m (11,500 ft), there are forests of Himalayan pine, cedar, birch and juniper.

YAK
(*Bos grunniens*) The flesh of the yak is as tender as beef but richer in nutrients, and yak milk has fed the Tibetans for millennia.

INTERNATIONAL BORDER
With such altitudes, the Himalayas is a natural border region spanning five countries: India, Pakistan, Nepal, Bhutan and China. This means that many of the main peaks have international approaches: Everest, Lhotse, Makalu and Cho Oyu, for example, can be reached from the Chinese side or the Nepali side.

ASIAN SPRING
Almost all the great rivers of south Asia have their sources in the lakes and springs of the Himalayas: the Indus has its source on the eastern side of the range and flows into the Arabian Sea, while the Ganges and the Brahmaputra also rise on the northern side, but release their waters into the Gulf of Bengal in the Indian Ocean. The Irrawaddy, Burma's great river, rises on the eastern side of the Himalayas and flows into the Andaman Sea, also in the Indian Ocean.

10,000
The number of attempts to climb Mount Everest in the last 50 years. 3,000 attempts were successful, leading about 2,000 people to the top.

Mount Everest
The highest mountain on Earth was conquered for the first time in 1953 (Edmund Hillary and Tenzing Norgay) and since then it has been climbed more than 3,000 times. Everest has claimed the lives of more than 200 people.

Edmund Hillary

Tenzing Norgay

Summit

Camp IX

Camp VIII

Camp VII

Camp VI

Camp V

Camp IV

Camp III

Camp II

Base Camp

SOUTH COL ROUTE

Salar de Uyuni

Located 11,995 ft (3,700 m) above sea level, and with a surface area of 4,674 sq miles (12,106 km²), in addition to being a huge salt flat in south-western Bolivia, Salar de Uyuni is one of the largest lithium reserves in the world. It was formed by the evaporation of ancient seas that bathed the American continent in times gone by. It comprises around 11 layers of salt, the thickness of which varies between 6.5 and 33 feet (2 and 10 metres).

LOCATION

▶ **References**

- 1 Salar de Uyuni
- 2 Salar de Coipasa
- Paleolake Tauca -15,000 years ago-
- Salt outcrops

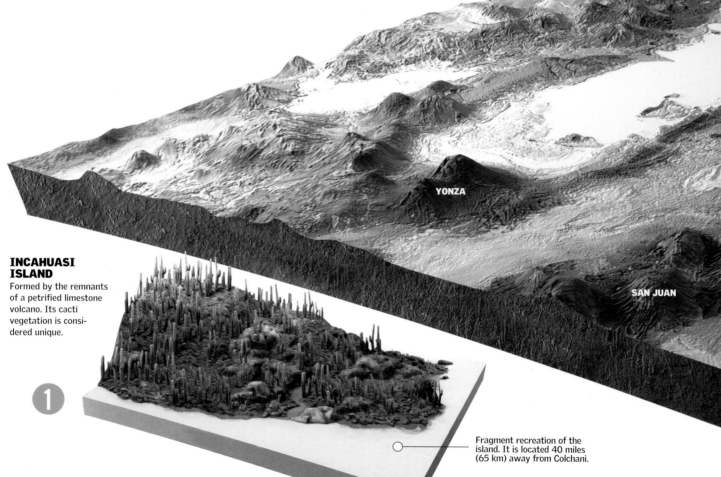

YONZA

SAN JUAN

INCAHUASI ISLAND

Formed by the remnants of a petrified limestone volcano. Its cacti vegetation is considered unique.

1

Fragment recreation of the island. It is located 40 miles (65 km) away from Colchani.

HOW SALAR DE UYUNI WAS FORMED

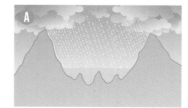

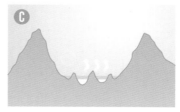

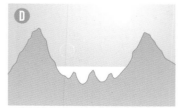

A Rain and melting freshwaters flooded the Bolivian highlands and formed a salt-water lake (the water contained salt as during the formation of the Andes, sea salt from previous layers rose to the surface and, when the area flooded, it dissolved in water).

B The sun and wind resulted in the evaporation of the water; the salt became separated and was left at the lower layer.

C New rainwater flooded the area once again.

D Again, the natural agents evaporated the water that flooded the highlands and salt continued to accumulate with each process of evaporation.

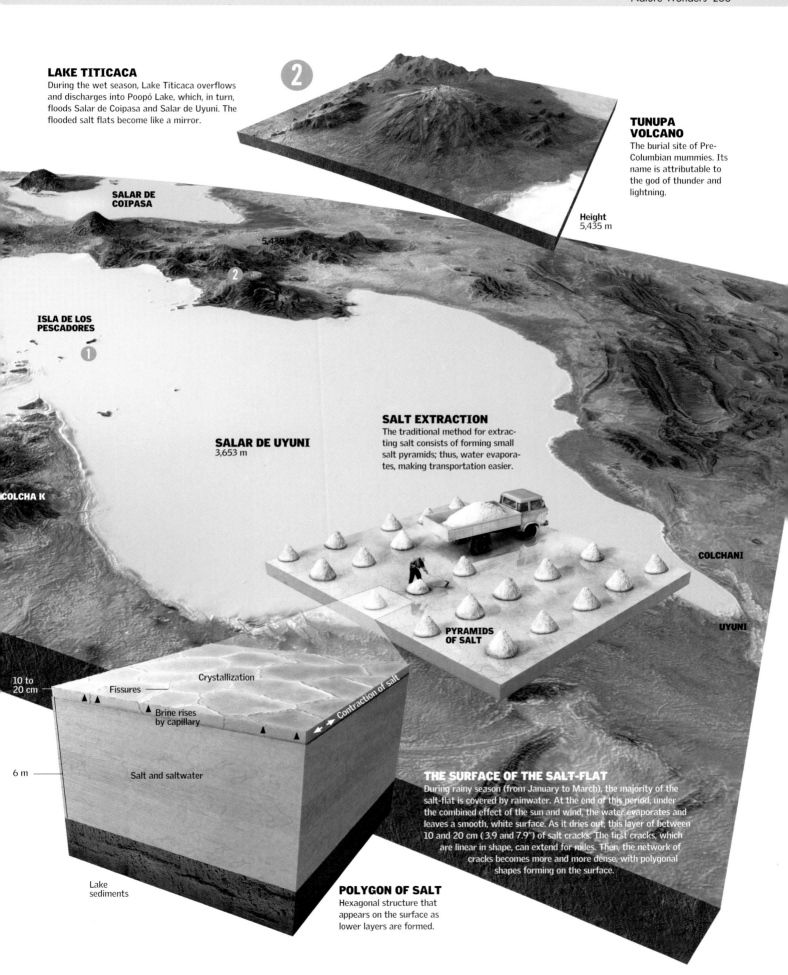

LAKE TITICACA

During the wet season, Lake Titicaca overflows and discharges into Poopó Lake, which, in turn, floods Salar de Coipasa and Salar de Uyuni. The flooded salt flats become like a mirror.

TUNUPA VOLCANO

The burial site of Pre-Columbian mummies. Its name is attributable to the god of thunder and lightning.

Height 5,435 m

SALAR DE COIPASA

5,436 m

ISLA DE LOS PESCADORES

SALAR DE UYUNI
3,653 m

SALT EXTRACTION

The traditional method for extracting salt consists of forming small salt pyramids; thus, water evaporates, making transportation easier.

COLCHA K

COLCHANI

UYUNI

PYRAMIDS OF SALT

10 to 20 cm

Crystallization

Fissures

Brine rises by capillary

Contraction of salt

6 m

Salt and saltwater

THE SURFACE OF THE SALT-FLAT

During rainy season (from January to March), the majority of the salt-flat is covered by rainwater. At the end of this period, under the combined effect of the sun and wind, the water evaporates and leaves a smooth, white surface. As it dries out, this layer of between 10 and 20 cm (3.9 and 7.9") of salt cracks. The first cracks, which are linear in shape, can extend for miles. Then, the network of cracks becomes more and more dense, with polygonal shapes forming on the surface.

Lake sediments

POLYGON OF SALT

Hexagonal structure that appears on the surface as lower layers are formed.

The Great Rift Valley

An immense gash in the Earth's surface, the world's largest rift system stretches 6,000 km from the Red Sea down to Lake Malawi. Up to 75 km wide in places, it's cradled by a series of steep cliffs, rising from the valley floor to the top of the highest escarpments up to 1.6 km above.

A complex Rift System

The Great Rift Valley is not one single feature but a series of linked faults: the Ethiopian Rift Valley, and the western and eastern branches of the East African Rift.

RED SEA

AFRICA

SUDAN

Nubian Plate

ERITR

ETHIOPIA

CENTRAL
AFRICAN
REPUBLIC

South Sudan

DEMOCRATIC
REPUBLIC OF
THE CONGO

*LAKE
ALBERT*

UGANDA

△ Mount
Moroto

*LAKE
TURKANA*

KEN

Mount Baker △

EAST PLATE

WEST
PLATE

RUANDA

BURUNDI

*LAKE
VICTORIA*

TANZANIA

*LAKE
TANGANYIKA*

*LAKE
NYASA*

LAKES
Africa's largest lake, Lake Victoria, sits between the western and eastern branches of the East African Rift. The two branches are thought to have developed because a thick, ancient chunk of metamorphic rock, called a craton, under the lake was too hard for the faults to penetrate.

HOW WAS THE GREAT RIFT VALLEY FORMED?

1 LIFTING THE CRUST
A hot mass of mantle rises, producing a ballooning effect on the crust's surface.

2 FORMATION OF THE RIFT
When the tectonic plates move, the continent experiences a great level of tension. It cracks and magma is allowed to escape, creating a great valley that experiences volcanic activity.

3 FROM THE VALLEY TO THE SEA
As the tension continues, it breaks the continental crust, a phenomenon known as rift. Between the continental blocks, oceanic crust is formed, with ramps in the middle. This is the case of the modern-day Red Sea.

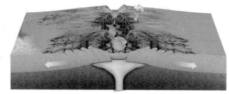

MAP

Showing tectonic plate boundaries, outlines of the elevation highs demonstrating the thermal bulges and large lakes of East Africa. Also, rift segment names for the East African Rift System. Smaller segments are sometimes given their own names, and the names given to the main rift segments change depending on the source.

▶ References

⌐ΛΛ⌐	Plates limit
··········	Plates limit development
☐☐☐☐☐☐☐☐	Dome limits
▬	African Rift boundary
▬	Ethiopian Rift boundary
🌲	Ngorongoro Reserve

ETHIOPIAN RIFT

The elevated flow of heat from inside the valley suggests that this geological fracture may have been caused by the so-called "hot spot", generating a plateau and the highlands in modern-day north-eastern Ethiopia and Eritrea. As the plateau rises, the immense pressure breaks the crust until it splits into three fractures, known as the Triple Junction: one is located in the Red Sea, another in the Gulf of Aden and the third through Ethiopia.

Arabian Plate

Plates boundary

Somali Plate

SOMALIA

ARABIAN SEA

All along the Great Rift Valley, blocks of rock slip downwards at a rate of 1mm a year to create 'grabens', which form the valley floor.

SAUDI ARABIA

AFRICA

YEMEN

Oceanic crust

Thinned crust

Flood basants

Oceanic crust

AFRICA

Monduli

Kitumbeine

Kerimasi   Lolmalasin

LAKE EMPAKAII

Oldoinyo Lengai Volcano

Empakaii Crater

Cráter Ngorongoro

LAKE MANYARA

Olmoti Crater

LAKE MAGADI

TANZANIA

Oldeani

Ngorongoro Reserve

Lemagarut

LAKE EYASI

Continental crust

Partial melting

Mantle plume

Continental crust

Lithosphere

Asthenosphere

NGORONGORO CONSERVATION AREA

The geologic forces that created this rift valley are still in motion, most visibly at the Ngorongoro Reserve.

1 The history of Mount Ngorongoro started around twenty or thirty million years ago, when the Great Rift Valley was formed. The molten material that rose through the faults resulted in the formation of a high number of volcanoes throughout the region, known as the Ngorongoro Highlands.

2 About five million years ago, Ngorongoro reached its maximum height, 4,570 m; but two million years later, geological activity developed deviated eastward, and consequently the volcanoes of the Crater Highlands became extinct.

3 This volcano awoke suddenly, accompanied by terrible explosions that hurled rocks and volcanic ash. A large underlying magma chamber was created, and the cone was sinking and introducing into the hollow chamber. This collapse caused the great Ngorongoro caldera: a gigantic crater-shaped depression.

The Grand Canyon

The Grand Canyon is not only one of the largest landscapes on Earth, it is an open book that demonstrates planet Earth's geological history. This marvel follows the course of the Colorado River in Arizona, over 277 mi (446 km) and forms part of the Colorado National Park.

CALIFORNIAN CONDOR
Gymnogyps californianus
After becoming almost extinct at the end of the last century the Grand Canyon was repopulated with condors raised in captivity.

CLEAR LAYERS
These layers are formed from Coconino sandstone and sea sediments (Toroweap), and have traces of vertebrates.

RED LAYERS
The three primary layers are Temple Butte and Redwall, both limestone rocks which contain a large number of fossils, and the Supai group of red sandstones.

GREYISH LAYERS
The Tonto group, formed 500 million years ago, has various horizontal layers with sandstone and limestone shale.

DARK LAYERS
In the deepest layers, which continue to be eroded by the Colorado River, there are rocks approaching two thousand million years old. When they were created, they were subject to very high temperatures.

COYOTE
Canis latrans
Lives close to the banks of the river, but it is difficult to see. It is more common to hear it howling.

Flora and Fauna

The Grand Canyon National Park has one of the richest ecosystems in the United States. It has all types of forest and desert, in addition to some 350 species of birds, and another 150 of mammals, reptiles, amphibians and fish.

1 mile
(1.600 m)

is the canyon's maximum depth. Its most narrow point from bank to bank is 0.5 miles (800 m).

AN INCREDIBLE VIEW

The Skywalk, built in 2007 with the approval of the Hualapai tribe, is a large circular platform with a glass floor which projects 20 m (66 ft) outwards and offers spectacular views of the Grand Canyon.

BARREL SHAPED CACTUS
Ferocactus cylindraceus.

UTAH JUNIPER
Juniperus osteosperma
This type of tree, together with the pinyon pine, is found predominantly in the extreme south of the Grand Canyon.

RAVEN
Corvus corax.

COUGAR
Puma concolor.

MOUNTAIN COTTONTAIL
Sylvilagus nuttallii.

MARGARITA FLEABANE
Erigeron divergens.

BEAVER TAIL CACTUS
Opuntia basilaris.

Iguazú Falls

Stretching 1.86 mi (3 km) across the border between Argentina
and Brazil, Iguazú Falls is one of the widest waterfalls in the
world. Its Guaraní name is Iguassu, meaning 'Big Water'. It is a
spectacular display of sound and fury.

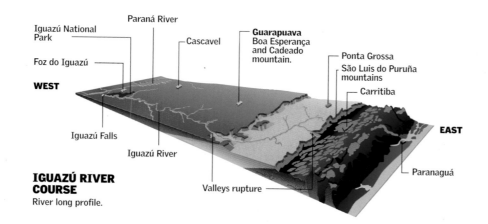

IGUAZÚ RIVER COURSE
River long profile.

DEVIL'S THROAT
Aerial view of the Iguazu River's spectacular fall into
the deep canyon. The sheer volume of the Iguazu
River falling 262 ft (80 m) causes a deafening roar.

THE ORIGIN
Iguazú Falls were originated
about 200,000 years ago in a
place called 'Three Borders', the
boundary between Argentina,
Paraguay and Brazil, due to a
fault. There, Iguazu and Parana
rivers come together. The fault
caused the mouth of the Iguazu
River to become a 262 ft
(80 m) high waterfall.

Paraná River

Iguazú River

Fault

PROFILE OF OUTPOURINGS OF BASALT AT THE FALLS

The staggered form of these falls is due to the structure of the three basalt spills that form it.

EACH OUTPOURING OF BASALT

When it has a thickness greater than 50 ft (15 m), it has three very well individualized portions: the upper, the middle and the lower.

Outpouring contact

580 ft (177 m)
570 ft (174 m) — Upper outpouring section
548 ft (167 m)
535 ft (163 m) — Central outpouring section — Upper outpouring
— Basal outpouring section
469 ft (143 m)
462 ft (141 m)
449 ft (137 m) — Upper outpouring section
436 ft (133 m)
— Central outpouring section — Middle outpouring
23 ft (7 m) — Basal outpouring section
3 ft (1 m) — Lower outpouring

THE HYDROLOGICAL ACTION

1. The force of water slowly wears away the soft rocks.

2. The hard rock (basalt), above the soft, begins to fracture and fall.

3. The remains of basalt, clay, lime, sand and gravel falling form the riverbed.

4. The cascade retracts due to the strength of the whirling water.

5. Meanwhile, the river pushes the debris.

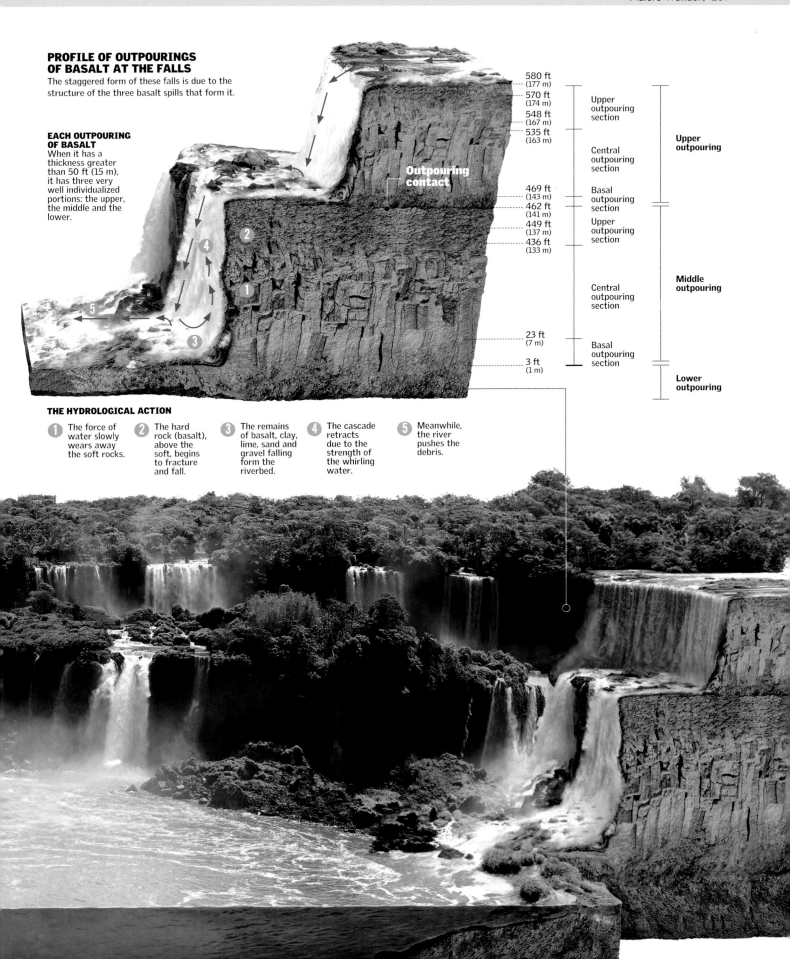

The Antarctic

The Antarctic, with temperatures of up to -158 °F (-70 °C) and winds of 124 mi/h (200 km/h), is a huge white continent and the coldest and windiest habitat of the globe. On its thick layer of ice, there are only algae, moss and lichen, especially in shore areas. However, plankton is abundant, due to the presence of ocean currents, rich in nutrients, and hydrothermal gaps.

Extreme survival

Many animals live in the shores, where they rest and reproduce, but in order to feed they have to enter the sea. Penguins, sole fixed residents of the Antarctic, bear the frigid Antarctic waters with their thick feathers and their abundant fat tissue. In turn, insects, such as the phlebotomus, count with anti-freezing proteins in their body fluids that prevent the formation of ice crystals in their cells.

WANDERING ALBATROSS
Diomedea exulans.
These sea birds grow up to 53 in (134 cm) in length and have a wing span of some 11 ft (3.5 m), making them the birds with the largest wingspan in the world. They eat squid, fish and the waste from fishing vessels.

CRABEATER SEAL
Lobodon carcinophagus.
They live in colonies of up to 50 million individuals. They mainly eat krill, but also fish and penguins. An adult seal can weigh between 397 and 507 lb (180 and 230 kg) and grow up to 8 ft (2.5 m) in length.

ELEPHANT SEAL
Mirounga leonina.

CHINSTRAP PENGUIN
Pygoscelis antarcticus.

FLOATING ICE

The extensive ice layer that covers the ocean, with an approximate thickness of 6.5 ft (2 m), is broken into smaller fragments of ice due to the wind and waves. These pieces of ice float on the ocean cover, since they have lesser density than liquid water, and they gradually disintegrate with the arrival of spring and summer.

ANTARCTIC PRION
Pachyptila desolata.

Snow
Covers the surface.

Firn
The upper layers "squash" the layers below. The snow is therefore compacted, turning it into firn.

Slopes
Made of ice, covering 44% of the coasts.

Tabular icebergs
Plates that can measure up to several miles in size, which tend to float adrift.

BLACK ROCKCOD
Notothenia coriiceps.

Continental ice
The deeper layers come under the most weight and therefore become more compact. Therefore, the firn turns into ice.

Under water
Two-thirds parts of an iceberg is submerged.

KRILL
Euphausia superba.
Huge schools of krill make their appearance during the summer. Its major customers are the whales that migrate to the Antarctic. They have special plates in their jaws that act as filters to catch.

5,400,000 sq mi
(14,000,000 km²)

is the approximate area of Antarctica (mainland plus surrounding islands).

ANIMAL-DRAWN TRANSPORT
Until the appearance of vehicles in the 20th century, a sledge, pulled by dogs from Greenland, was the only method of transport used on the continent.

KILLER WHALE
Orcinus orca.
Killer whales hunt warm-blooded animals in packs. They locate their prey on the ice floes and, with their sharp teeth, tear at the flesh.

SPERM WHALE
Physeter macrocephalus.
Sperm whales are the largest living carnivores on the planet. They can be found throughout the world except the Arctic and usually go on long migratory journeys. The males travel to the Antarctic to find food.

LEOPARD SEAL
Hydrurga leptonyx.

GENTOO PENGUIN
Pygoscelis papua.

ANTARCTIC MINKE WHALE
Balaenoptera bonaerensis.
The minke whale is the smallest of the cetaceans. It feeds using its baleen, bristles which filter the water and retain food, mainly krill.

EMPEROR PENGUIN
Aptenodytes forsteri.
With a 3.3 ft (1 m) height and 99 lb (45 kg) weight, this species is characterized for being the largest, in addition to showing the best colors. Unlike other penguins, females place their eggs on the frozen ocean, where large colonies are formed.

Treaty

Home to the South Pole, this continent is governed by the Antarctic Treaty, which has 50 signatory countries, although only 28 have the right to vote.

The Amazon

The basin of the longest and mightiest river in the world is home to an extensive tropical rainforest, and is populated by a range of flora and fauna not found anywhere else on Earth. The great Amazon rainforest spans nine South American countries, although the largest part of the forest, 60 percent, is found in Brazil.

THE AMAZON BASIN

Like a great country

It extends over more than 2.3 million mi² (6 million km²), which would make the Amazon the seventh largest country in the world. Located in South America, between the Andes mountain range and the Atlantic Ocean, it occupies almost the entire basin of the Amazon, the largest and mightiest river on the planet. It has an extensive area of lowlands and plains covering the northern half of Brazil, the south of Colombia, the east of Ecuador and Peru, and the north of Bolivia.

80%

of medication used in western medicine has been synthesized based on products extracted from plants, fungi and animals from the Amazon basin.

THE RIVER
It is the river of the records. Born on Nevado de Mismi (Peru) at 18,372 ft (5.600 m) above sea level and after traveling over 3,975 mi (6.400 km) it flows into the Atlantic, forming a broad estuary about 150 mi (240 km) away.

A paradise of biodiversity

The Amazon has the largest reserve of biological diversity on the planet. The region represents 40 percent of the tropical rainforests in the world and is home to half of all known animal and vegetable species. This is in contrast to the above mentioned boreal Taiga, in which just one woodland species—usually a conifer—monopolizes enormous areas of land. In just one hectare of Amazon rainforest, 300 different species of tree can be identified, and up to 650 species of beetle can be found on each one.

BIRDS
1,000
Species

FISH
3,000
Species

INSECTS
10 mill.
Species

REPTILS
550
Species

MAMMALS
350
Species

AMAZON FOREST

The height and the leafiness of the trees allow life to be distributed over various levels, from the ground, where very little light reaches, to the canopy in the treetops, where the intense sunlight provokes the concentration of a great variety of fauna which never descend to lower levels.

OVERSTOREY

This is the highest part of the forest, to which the largest trees extend, up to 230 ft (70 m). It receives a lot of sun and eagles, parrots, bats, butterflies and monkeys live in it.

KING VULTURE
Sarcoramphus papa.

SPIDER MONKEY
Ateles belzebuth.

SQUIRREL MONKEY
Saimiri sciureus.

PAVONINE QUETZAL
Pharomachrus pavoninus.

HOWLER MONKEY
Alouatta palliata.

CANOPY

Extending to a height of between 98 and 164 ft (30 and 50 m), more than 50 percent of the plants grow here and between 70 and 90 percent of the species live here.

EMERALD TREE BOA
Corallus caninus.

COLLARED ANTEATER
Tamandua tetradactyla.

RING-TAILED COATI
Nasua nasua.

TOCO TOUCAN
Ramphastos toco.

SPECTACLED OWL
Pulsatrix perspicillata.

UNDERSTOREY

The upper layers of vegetation prevent much light from reaching the bottom. Here we find orchids, lichens and bracken and many snakes and frogs.

CRIMSON-CRESTED WOODPECKER
Campephilus menlanoleucus.

ANTEATER
Myrmeciphaga tridactyla.

GROUND LEVEL

Less than 2 percent of the sun's rays reach the ground. It is the habitat of the largest animals as well as thousands of invertebrates.

Kilimanjaro

At 19,331 ft (5.892 m) above sea level, Mount Kilimanjaro is the highest point in Africa, the king of the African savannah, over which it presides with power and grandeur. This colossus is located in the north-east of Tanzania and forms part of the nature park named after it.

Three in one

Kilimanjaro is an enormous volcanic mountain made up of three aligned cones: Shira, to the west, is 13,000 ft (3.962 m) high; Mawenzi, to the east, reaches 16,900 ft (5.149 m); and Kibo, in the center, is the highest of the three at 19,330 ft (5.892 m), and forms the overall peak, known as Uhuru and was climbed for the first time in 1889.

ENVIRONMENTS
Due to its altitude, Kilimanjaro has different environments with their own characteristics.

- Snow environment
- Afro-alpine environment
- Mattoral, plain and scrub
- Cloud forest
- Rainforest
- Savannah and plantations

SHIRA CLIFFS

Extended area

SHIRA CONE
The oldest and most eroded of the three cones of Kilimanjaro, it measures 13,000 ft (3.962 m).

GLACIERS IN RETREAT
If the current tendency, provoked by global warming, continues, the mountain may lose all its glaciers before 2030.

1990

2010

UNDER PROTECTION
The Kilimanjaro National Park occupies an area of 75,353 hectares (186,201 acres) and includes both the volcano and the mountain forests and part of the savannah that surrounds it.

FAUNA
On the surrounding plains there are large mammals such as lions, zebras and elephants.

NOT SO ALONE

In reality, Kilimanjaro forms part of a chain of volcanoes located on the eastern margin of the Great Rift Valley. The image shows its lineup with an extensive volcanic chain oriented from east to west.

Monduli

Meru

Crater of Ngurdoto

Shira

Kibo

Mawenzi

THE MASAI
The north and west of the volcano are occupied by the Maasai, a pastoral people, one of whose most ancient traditions is to start controlled fires to regenerate the earth with the carbonized vegetation.

CONE OF KIBO
This is the youngest of the cones. Its profile is surrounded by gentle slopes. It still releases volcanic gases.

THE SADDLE
This valley, located between the cones of Kibo and Mawenzi, is located at about 11,800 ft (3.600 m) above sea level, and contains the largest area of mountain tundra vegetation in Africa.

REUSCH CRATER

BREACH WALL
This is a wall that borders a deep fissure in the mountain, the product of an ancient landslide.

UHURU PEAK OR KIBO
A sign in English welcomes climbers at the top of Kilimanjaro.

60,000

is the estimated number of people visiting Kilimanjaro each year.

MAWENZI CONE
At 16,900 ft (5.149 m) it is the third highest peak in Africa. Although it is only 3.7 mi (6 km) away from the cone of Kibo, it is completely different, with a scarped, abrupt profile.

Cappadocia

The formation of the mountain chain of Taurus, 60 million years ago, created a zone of ravines and depressions in central Anatolia, which was later covered by lava from different volcanoes in the region. For centuries, extreme changes in temperature, rainfall and weather were sculpting an incredible landscape of rocks, chimneys and caves.

Nature's sculptor

The 300 sq km region of Cappadocia sits on a high plateau in the center of Turkey, far from the balmy influence of the sea, so summers are very hot and winters very cold. These extremes expand and contract rock time and again. Gradually, the softer tuff was worn away by this weathering and water erosion, leaving distinctive geological formations, such as the towering 'fairy chimneys'.

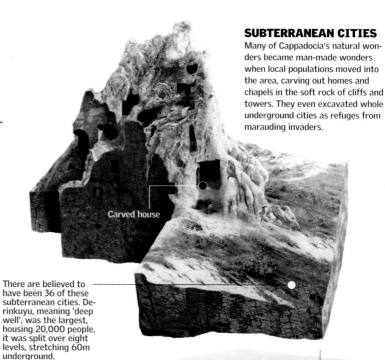

SUBTERRANEAN CITIES
Many of Cappadocia's natural wonders became man-made wonders when local populations moved into the area, carving out homes and chapels in the soft rock of cliffs and towers. They even excavated whole underground cities as refuges from marauding invaders.

Carved house

There are believed to have been 36 of these subterranean cities. Derinkuyu, meaning 'deep well', was the largest, housing 20,000 people, it was split over eight levels, stretching 60m underground.

FORMS OF THE CHIMNEYS
The peculiar morphology of these geological formations is due to the vertical accumulation of sedimentary material on which rests a layer of rock more resistant to erosion, the dome or cap.

Ignimbrite
Sediments

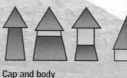

Only cap   Only body   Cap and body

Cap
Neck
Body

Fairy chimneys can reach heights of 131 ft (40 m).

BEWITCHED LANDSCAPE

The locals call these 'fairy chimneys', as the ancients belie-ved that fairies living underground came up to the surface through these pillars.

36

underground cities exist in the area of Cappadocia. They are narrow, damp places where its first inhabitants would hide while under attack.

HOW CHIMNEYS WERE FORMED

Geological effects caused changes in the terrain. Millions of years of erosion created fanciful shapes.

1 Eruption of volcanoes a million centuries ago, during the Miocene and Pliocene.

2 Soft rock layer called tuff, result of the first eruptions.

3 Following eruptions created a layer of basalt, a harder mate-rial, which avoided the erosion of the interior tuff.

4 Finally, after millions of years, sand and wind eroded the tuff. This process gave rise to the chimneys.

INDEX

INDEX

A

accretion, 22-23
abdomen, 132-133, 152-153, 166-167, 172-173, 175-176, 196-197
acorn, 144-145, 180-181
aerenchyma, 184-185
air, 14-15, 16-17, 24-25, 32-33, 38-39, 40-41, 84-85, 86-87, 94-95, 106-107, 108-109, 110-111, 112-113, 114-115, 116-117, 118-119, 120-121, 122-123, 124-125, 126-127, 228-229
albedo, 98-99, 100-101, 106-107
alcohol, 222-223
algae, 84-85, 86-87, 96-97, 182-183, 202-203, 204-205, 206-207, 241-242
  types, 202-203:
    chlorophytes
    phaeophytes
    rhodophytes
alligator, 166-167
allergenics, 220-221
Amazon, 242-243
amethyst, 50-51
ammonites, 28-29
amphibian, 164-165
  orders:
    Anura
    Apoda
    Urodela
androecium, 194-195
anemone, 96-97
Animalia (animals), 82-83
annulus, 210-211, 212-213
Antarctic, 240-241
  Antarctic Treaty
anther, 194-195
antheridium, 204-205, 210-211, 212-213
antherozoid, 212-213
anticyclone, 110-111
apatite, 48-49
aphelion, 16-17
aphotic, 94-95
apothecia, 208-209
appendage, 172-173
arachnid, 174-175
  types:

Acari
Amblypygi
  scorpions
  spiders
Archaea, 82-83
Archean Eon, 22-23
archegonium, 210-211, 212-213
Arctic, 130-131
Armero (Colombia), 64-65
ascocarp, 214-215
ascomycetes, 218-219
ascospores, 214-215, 218-219, 220-221, 222-223
ascus, 214-215, 222-223
ash (pyroclastic product), 62-63, 106-107
aspergillosis, 220-221
aspergillum, 220-221
atmosphere, 14-15, 24-25, 98-99, 106-107, 110-111, 122-123
  layers, 24-25:
    troposphere
    stratosphere
    mesosphere
    thermosphere
    exosphere
Augustine, mount, 62-63
Autumn, 186-187
axilla, 190-191

B

barnacle, 172-173
basidia, 214-215, 216-217
basidiocarp, 216-217
basdiomycetes, 216-217
bat, 148-149, 156-157
batholiths, 24-25
bear, 90-91, 130-131, 132-133
bedrock, 84-85
bee, 196-197
beeches, 178-179
beer, 222-223
bill, 152-153
biodiversity, 88-89, 90-91, 242-243
bioma, 92-93
  land biomes:
    deserts
    forests
    grasslands

mountains
  polar regions
biorhythm, 144-145
biosphere, 14-15, 106-107
biotope, 86-87
bird, 152-153, 154-155, 156-157
blackberry, 198-199
bladder, 158-159
bladderwort, 184-185
blowhole, 146-147
boa, 166-167, 168-169, 242-243
bomb (pyroclastic product), 62-63
Bora Bora, 96-97
brachiopods, 28-29
bradycardia, 146-147
bread, 220-221, 222-223
bryophites, 210-211
bud, 138-139, 180-181, 222-223
bulla, 138-139

C

caecilian, 164-165
caiman, 166-167
calcite, 48-49
caldera (volcano), 60-61
Caldera Blanca (Canary Islands), 60-61
Calvin cycle, 182-183
cambium, 190-191
camel, 226-227
camouflage, 196-197
cap, 216-217
Cappadocia, 246-247
  Fairy chimneys
  Subterranean cities
capsule, 210-211
carbon, 86-87, 182-183
  carbon cycle, 86-87, 98-99
  carbon dioxide, 98-99, 182-183, 222-223
carbonization, 54-55
carpel, 194-195
cattails, 184-185
caudicle, 196-197
cellulose, 182-183
cephalotorax, 172-173, 174-175

cetacean, 136-137
Challenger Deep, 94-95
Chapel of St. Michael (France), 60-61
chelicerae, 174-175
chestnut, 144-145, 178-179
chipmunk, 130-131
Chiroptera, 136-137, 148-149
chlorine, 98-99
chlorofluorocarbons, 98-99, 102-103
chlorophyll, 182-183
chlorophytes, 202-203
chloroplast, 182-183
chondrichtyes, 158-159
class, 82-83
climate, 90-91, 100-101, 106-107, 108-109
  climate change, 100-101
  climate zones, 108-109
  Köppen climate classification, 108-109
  temperature, 108-109
  types:
    arctic, 90-91
    subarctic, 90-91
    temperate, 90-91, 108-109
    tropical, 90-91, 108-109
climatology, 106-107, 108-109, 110-111, 112-113, 114-115, 116-117, 118-119, 120-121, 122-123, 124-125
cloud, 14-15, 32-33, 116-117, 118-119, 126-127
  formation, 116-117
  types, 116-117:
    high clouds:
      cirrus
      cirrocumulus
      cirrostratus
    medium clouds:
      altocumulus
      cumulonimbus
      altostratus
    low clouds:
      cumulus
      stratocumulus
      nimbostratus,
      stratus
clown fish, 91-92, 96-97
coal, 38-39, 54-55
  phases and types, 54-55:

vegetation
peat
lignite
coal
anthracite
cobra, 226-227
codium, 206-207
Coleoidea, 170-171
condensation, 14-15, 32-33, 94-95
conidia, 220-221
conidiophores, 214-215, 220-221
conifer, 192-193
continent, 22-23, 26-27, 28-29, 30-31
creation of continents, 22-23
continental drift, 26-27
continental plates, 26-27, 28-29, 30-31
convergent boundary, 26-27
divergent boundary, 26-27
orogeny, 28-29, 30-31
faults, 30-31
mountain's formation, 28-29
tectonic movement, 26-27, 28-29, 30-31
coral, 96-97
core (Earth), 18-19, 24-25, 70-71
inner core, 18-19, 24-25, 70-71
outer core, 18-19, 24-25, 70-71
Coriolis effect, 34-35, 110-111, 112-113, 114-115
corolla, 194-195
cortex, 208-209
corundum, 48-49
cotyledon, 186-187
cow, 142-143
crab, 172-173
cranium, 130-131
crocodile, 166-167
crustacean, 172-173
anatomy
class:
Copepoda
Malacostraca
order:
Isopoda
cryosphere, 106-107

currents, 26-27, 34-35, 110-111, 114-115, 126-127
convection currents, 26-27
ocean currents, 34-35, 110-111, 114-115
air currents, 110-111, 126-127
cuticle, 178-179, 190-191, 216-217
cyclone, 110-111, 126-127

D

Darwin, Charles, 96-97
Dead Sea, 94-95
Death Valley, 228-229
decomposers, 86-87
deforestation, 100-101
dentititio, 130-131
desert, 90-91, 92-93, 108-109, 226-227
desert hedgehog, 226-227
desert rose, 226-227
destroying angel, 218-219
dew, 120-121
diamond, 48-49, 50-51
chemistry, 50-51
phases of production:
extraction
cutting and carving
polishing
precious stones
semiprecious stones
dicotyledon, 188-189
digit, 156-157
digitigrade, 136-137
dik-dik, 92-93
diploid, 210-211, 222-223
diving, 146-147
dog, 138-139
domain, 82-83
dolphin, 82-83, 130-131, 150-151
dromedary, 226-227
droplet, 31-32, 94-95, 106-107, 118-119, 120-121
dune, 226-227, 228-229

E

ear, 150-151, 154-155
Earth, 14-15, 16-17, 18-19,

20-21, 22-23, 24-25, 40-41, 42-43, 90-91, 98-99, 100-101, 236-237
atmosphere, 24-25
axis, 14-15
characteristics, 14-15
continent, 22-23
crust, 18-19, 24-25
geographic coordinate, 16-17
history, 22-23
internal layers, 24-25:
lithosphere
asthenosphere
upper mantle
lower mantle
magnetism, 18-19, 100-101
mantle, 18-19, 24-25
movements, 14-15, 16-17
rotation, 14-15, 16-17
revolution, 14-15, 16-17
nutation, 16-17
precession, 16-17
orbit, 14-15, 16-17, 100-101
orogeny, 28-29
surface, 24-25, 40-41, 42-43, 110-111, 236-237
surface's changes, 40-41
erosion, 40-41, 42-43
folding, 40-41
fracture, 40-41
magmatism, 40-41
metamorphism, 40-41
weathering, 40-41
earthquake, 68-69, 70-71, 72-73, 74-75, 76-77, 78-79
description, 68-69
epicenter, 68-69
hypocenter, 68-69, 70-71
measuring, 72-73
EMS 98 Scale
Mercalli Scale
Richter Scale
origin, 68-69
phases, 68-69
foreshock
aftershock
seismic energy, 68-69, 70-71, 72-73, 74-75
seismic waves, 70-71
world seismic areas, 78-79
earthworm, 84-85

echolocation, 150-151
eclipse, 20-21
lunar eclipse
solar eclipse
ecosystem, 86-87, 88-89, 94-95
aquatic ecosystem, 94-95
fresh water, 94-95
open ocean, 94-95
sea coasts, 94-95
ectothermic, 166-167
eelgrass, 184-185
Ekman spiral, 34-35
elephant seal, 130-131
emerald, 50-51
endosperm, 186-187
energy, 38-39, 86-87
energy flows, 86-87
nonrenewable energy, 38-39
fossil chemical energy
coal
natural gas
petroleum
nuclear energy
renewable energy, 38-39
biodigester
biofuel energy
geothermal energy
hydroelectric energy
hydrogen energy
solar energy
tidal energy
wind energy
sources of energy, 38-39
enzyme, 222-223
epidermis, 160-161, 184-185, 190-191, 192-193
epiphyte, 178-179
equinox, 16-17
erg, 226-227
ergot, 214-215, 218-219
ergotism, 218-219
erosion, 40-41, 42-43, 92-93
eolian processes, 40-41
Eubacteria, 82-83
evaporation, 14-15, 32-33, 94-95, 106-107
Everest, mount, 94-95, 230-231
exfoliation (mineral), 48-49
exoskeleton, 42-43, 96-97, 172-173, 174-175

exosphere, 24-25, 116-117
extremity, 136-137

F

family, 82-83
fat, 132-133
fault, 30-31, 78-79, 228-229
  types, 30-31:
    normal fault
    oblique-slip fault
    reverse fault
    strike-slip fault
  New Zealand fault, 78-79
feather, 152-153, 156-157
  types, 156-157:
    alula
    primaries
    primary coverts
    secondaries
    median wing coverts
    tertiaries
    greater wing coverts
feet, 152-153
feldspar, 46-47
feline, 136-137
fennec, 226-227
fern, 178-179, 212-213
fertilization, 194-195, 204-205, 210-211, 212-213
fig, 198-199
filament, 194-195, 212-213
fin, 150-151, 158-159
  anal fin, 158-159
  caudal fin, 150-151, 158-159
  dorsal fin, 150-151, 158-159
  pectoral fin, 150-151, 158-159
fish, 158-159, 160-161, 162-163
  anatomy, 158-159
  lateral line, 158-159
  scales, 160-161
  school, 162-163
  swimming, 162-163
  class, 158-159:
    Chondrichthyes
    Cyclostomata
    Osteichthyes
flour, 218-219
flower, 180-181, 194-195, 196-197
  clade 194-195:
    dicotyledon
    monodicotyledon

fluorite, 48-49
fly agaric, 216-217, 218-219
flying, 152-153, 156-157
fog, 120-121
  types:
    advection fog
    frontal fog
    inversion fog
    orographic fog
    radiation fog
fold, 68-69
foliaceous, 208-209
food, 86-87
foot, 134-135, 136-137
forest, 90-91, 92-93
  coniferous forest, 92-93
  deciduous forest, 92-93
  tropical forest, 92-93
Franklin, Benjamin, 122-123
freshwater, 32-33, 94-95
frog, 164-165
frond, 212-213
fruit, 198-199
  classification:
    dry fruits
    multiple fruits:
      aggregate fruit
      syconium
    simple fruits:
      berries
      drupes
      pomes
  description:
    endocarp
    mesocarp
    exocarp
fucus, 202-203, 204-205
Fuji, mount, 60-61
Fujita, Theodore, 124-125
fumarole, 36-37
Fungi, 82-83, 178-179, 214-215, 216-217, 218-219, 220-221
  division, 214-215:
    Ascomycota
    Basidiomycota
    Chytridiomycota
    Deuteromycota
    Zygomycota
  parasites, 216-217
  saprobes, 216-217

G

gametophyte, 210-211, 212-213
gamosepalous, 194-195
gastropod, 170-171
gazelle, 92-93
gem, 50-51, 222-223
gemation, 222-223
gemsbok, 226-227
gen, 88-89
genus, 82-83
geotropism, 188-189
gerbil, 226-227
gerenuk, 92-93
germination, 212-213
geyser, 36-37
gibberellins, 186-187
gills, 158-159, 216-217
giraffe, 92-93
gland, 32-33, 82-83, 130-131, 158-159, 164-165, 170-171, 172-173, 174-175, 194-195
  digestive gland, 170-171, 172-173
  mammary gland, 82-83, 130-131
  mucous gland, 164-165
  poisonous gland, 164-165
  salivary gland, 170-171, 174-175
  silk gland, 174-175
  sweat gland, 32-33, 130-131
  venom gland, 174-175
glucose, 190-191, 222-223
glucosys, 222-223
gnu, 92-93
gonidia, 208-209
gorilla, 130-131
Grand Canyon of the Colorado, 42-43, 236-237
granite, 46-47
grass, 178-179
grassland, 90-91, 92-93
gravity, 14-15, 18-19
greenhouse effect, 98-99, 106-107
  concentration CO_2, 98-99
  greenhouse gases, 98-99, 100-101
    carbon dioxide, 98-99
    chlorofluorocarbons, 98-99
    methane, 98-99

nitrogen oxides, 98-99
stratospheric ozone, 98-99
gynoecium, 194-195
gypsum, 48-49

H

habitat, 90-91
  climate factor, 90-91
  distribution, 90-91
  habitats of the world, 90-91
    coniferous forest
    coral reefs
    deserts
    grasslands
    savanna
    steppe
    mountains
    polar regions
    temperate forest
    tropical forest
hadal, 94-95
hail, 14-15, 32-33, 112-113, 116-117, 118-119, 122-123
hair, 130-131, 132-133, 208-209
hamada, 226-227
haploid, 210-211, 222-223
hazel dormouse, 144-145
hearing, 138-139
herbivore, 142-143
hibernation, 144-145, 148-149
Hillary, Edmund, 230-231
Himalayas, 28-29, 230-231
homeothermy, 130-131, 132-133
hoof, 134-135
hornwort, 184-185
horse, 134-135
Houston, 108-109
Howard, Luke, 116-117
hummingbird, 90-91, 152-153
humus, 84-85
hurricane, 126-127
  formation
  classification of damage
hydrosphere, 106-107
hymenium, 216-217
hyphae, 208-209, 214-215, 216-217, 220-221
hypocotyl, 186-187
hypodermis, 192-193

I

ice, 14-15, 28-29, 32-33, 62-63, 90-91, 94-95, 100-101, 106-107, 116-117, 118-119, 122-123, 132-133, 240-241
Iguazú Falls, 238-239
Ilamatepec, mount (El Salvador), 60-61
Indian Ocean, 76-77
indusium, 212-213
internode, 190-191
isobar, 110-111

J

jaw, 130-131
jet-lag, 16-17
jet stream, 110-111
Jupiter, 18-19

K

Kilauea, mount (Hawaii), 60-61, 62-63
Kilimanjaro, mount, 244-245
Kimberley Mine (South Africa), 50-51
kingdom, 82-83
kinship, 82-83
knotweed, 184-185
Köppen, Waldimir, 108-109

L

labellum, 196-197
lachenalia, 184-185
lahar, 64-65
lake, 14-15, 32-33, 34-35, 234-235
Lamellibranchiata, 170-171
lamprey, 158-159
lapilli, 62-63
lava, 62-63, 64-65
leaf, 178-179, 180-181, 182-183, 186-187, 192-193, 194-195
    compound leaves, 192-193
    simple leaves, 192-193
leptosporangium, 212-213
Lhasa, 108-109
lichen, 208-209
    types:
        foliaceous

fructicose
    corticolas
    saxicola
    terricola
light, 182-183
lightning, 122-123
    phases
    types:
        cloud-to-air
        cloud-to-cloud
        cloud-to-ground
lightning rod, 122-123
limb, 130-131
lindens, 178-179
lionfish, 96-97
lithosphere, 24-25, 106-107
lizard, 166-167
    Solomon island skink, 166-167
lobule, 196-197
louse, 172-173
lungs, 130-131, 132-133, 146-147, 150-151, 164-165, 166-167, 168-169, 170-171, 174-175

M

macrocystis, 206-207
magmatism, 40-41
magnetosphere, 18-19
Maka-O-Pulh (Hawaii), 62-63
mammal, 130-131
Manaus, 108-109
mandible, 150-151
mantle (Earth), 18-19, 24-25, 70-71
    upper mantle, 24-25
    lower mantle, 24-25
maple, 178-179, 192-193
marble, 52-53
Mars, 14-15, 14-15, 18-19
Masai, 244-245
Maula Ulu (Hawaii), 60-61
medulla, 208-209
meiosis, 210-211, 222-223
melon, 150-151
Mercalli, Giuseppe, 72-73
Mercury, 18-19
mesosphere, 24-25, 116-117
metabolism, 132-133, 146-147, 152-153, 222-223

metamorphism, 40-41
methane, 98-99
mica, 46-47
migration, 132-133
minerals, 46-47, 48-49
    density, 48-49
    exfoliation and fracture, 48-49
    hardness, 48-49
    Mohs scale, 48-49
    properties, 48-49
    structures, 46-47
mist, 120-121
mite, 174-175
mitochondria, 222-223
mitosis, 222-223
Mohs, Friedrich, 48-49
Mojave Desert, 228-229
mold, 214-215
mollusk, 170-171
    anatomy
    bivalve
        Lamellibranchiata
        Protobranchia
    cephalopod
        Coleoidea
        Nautiloidea
    gastropod
        Opisthobranchia
        Prosobranchia
moloch, 90-91
monocotyledon, 188-189
monsoon, 114-115
Moon, 14-15, 18-19, 20-21
Moscow, 108-109
moss, 178-179, 210-211
mouth, 146-147
mudslide, 64-65
mushroom, 214-215, 218-219
mycelium, 214-215, 216-217
mycotoxin, 218-219

N

nectar, 196-197
Neptune, 18-19
nest, 144-145
Nevado del Ruiz, mount, 64-65
neuron, 150-151
New Zealand, 68-69
nitrogen, 86-87, 98-99

node, 190-191
Norgay, Tenzing, 230-231
nut, 144-145
nutclam, 170-171
nutrient, 186-187, 188-189

O

oak, 178-179
oasis, 226-227
ocean, 14-15, 32-33, 34-35, 94-95
octopus, 82-83, 96-97, 170-171
odor, 196-197
omasum, 142-143
opal, 50-51
operculum, 210-211
orange, 198-199
orchid, 196-197
order, 82-83
orogeny, 28-29
    Alpine orogeny
    Caledonian orogeny
    Hercynian orogeny
orthoclase, 48-49
ostrich, 152-153, 156-157
osmosis, 188-189
ovary, 168-169, 194-195
oviparous, 166-167
ovule, 212-213
ovum, 204-205
oxygen, 146-147, 182-183
ozone, 98-99, 102-103
    ozone hole, 102-103
    ozone layer, 102-103, 106-107

P

Paleozoic Era, 22-23
parenchyma, 190-191
parrot feather, 184-185
patagiu, 148-149
pedipalp, 174-175
peel, 198-199
pegmatite, 52-53
penguin, 152-153
perch, 160-161
perianth, 194-195
perihelion, 16-17
perithecium, 218-219

permafrost, 84-85, 92-93
petal, 194-195
petroleum, 38-39, 54-55
  phaeophytes, 202-203
pheromone, 196-197
phialides, 220-221
phloem, 190-191, 192-193
photic, 94-95
photosynthesis, 178-179, 182-183, 192-193, 208-209
phylum, 82-83
piezoelectricity, 48-49
pincer, 172-173
pinnas, 212-213
pinnules, 212-213
Plantae (plants), 82-83, 178-179, 180-181, 182-183, 184-185, 186-187, 188-189, 190-191, 192-193, 194-195, 196-197, 198-199, 202-203, 204-205, 206-207, 208-209, 210-211, 212-213, 214-215, 216-217, 218-219, 220-221
  aquatic plants, 184-185
  fruit, 196-197
  growth, 186-187
  plant cells, 182-183
    Calvin cycle, 182-183
    chloroplast, 182-183
    stroma, 182-183, 218-219
    thylakoids, 182-183
    vacuole, 182-183, 222-223
  plant tissues, 182-183
plantigrade, 136-137
plate (orogeny), 26-27, 74-75, 78-79
  plate tectonic, 26-27
  rising plate, 74-75
  sinking plate, 74-75
  subduction zone, 26-27
plumage, 152-153
plumule, 186-187
plutons, 24-25
pneumatophores, 184-185
pole, 14-15, 18-19
pollen, 196-197
pollination, 196-197
poppy, 186-187
porphyra, 206-207
potato, 190-191
precipitation, 14-15, 32-33, 94-95, 106-107
pressure (weather), 110-111

Prosobranchia, 170-171
Proterozoic Eon, 22-23
prothallus, 212-213
Protista (protozoa), 82-83
pterodactyl, 152-153
pupil, 140-141
pyroelectricity, 48-49
python, 168-169

Q

quake, 64-65
quartz, 46-47, 48-49

R

rachis, 192-193, 212-213
radar, 148-149
radicle, 186-187
rain, 14-15, 32-33, 40-41, 58-59, 64-65, 84-85, 90-91, 92-93, 94-95, 104-105, 108-109, 112-113, 114-115, 118-119, 120-121, 122-123, 126-127, 212-213, 226-227, 232-233, 246-247
  acid rain, 37-38, 40-41
  rainforest, 88-89, 90-91, 108-109, 166-167, 208-209, 210-211, 242-243, 244-245
  rainwater, 58-59, 232-233
reef, 96-97
  coral reefs, 96-97
  barrier reef, 96-97
reg, 226-227
remige, 156-157
reptile, 166-167, 168-169
  order, 166-167:
    chelonians
    crocodiles
    Squamata
resin, 192-193
respiration, 146-147
retina, 140-141
rhizome, 212-213
rhodophytes, 202-203
rib, 168-169
Richter, Charles, 72-73
ricin, 208-209
ridge, 24-25, 78-79
Rift Valley, 234-235
ring, 66-67, 180-181, 216-217
  growth rings (tree), 180-181

Ring of fire (volcano), 66-67
river, 14-15, 32-33, 242-243
rocks, 44-45, 46-47, 50-51, 52-53, 54-55, 84-85
  color, 52-53
  cycle, 84-85
  fracture, 52-53
  igneous rock, 84-85
  metamorphic process, 44-45
  metamorphic rock, 44-45, 84-85
    fusion, 44-45
    gneiss, 44-45
    schist, 44-45
    slate, 44-45
  mineral composition, 46-47, 50-51, 52-53, 54-55, 84-85
    organic rocks, 54-55
    precious stones, 50-51
    sedimentary rock, 84-85
root, 178-179, 186-187, 188-189, 194-195, 212-213
  types, 188-189:
    branched
    fibrous
    napiform
    tabular
    taproot
    tuberous
ruby, 50-51
rumination, 142-143
rye, 218-219

S

Sahara, 226-227
sailfish, 162-163
Salar de Uyini, 232-233
salinity, 158-159
salmon, 160-161
salt, 94-95, 228-229, 232-233
San Andreas fault, 30-31
sandfish, 226-227
sapphire, 50-51
saprobes, 216-217
saprobia, 220-221
sapwood, 190-191
Saturn, 18-19
scale, 160-161, 168-19
scorpion, 174-175, 226-227
Scotland, 44-45
season, 14-15, 16-17, 34-35,

90-91, 180-181
seaweed, 203-204
seed, 180-181, 186-187, 198-199
sense, 138-139, 152-153, 134-135
sepal, 194-195
sepia, 170-171
shark, 82-83, 158-159, 160-161, 162-163
shelf, 24-25
shell, 172-173
shrimp, 172-173
sieve, 190-191
skeleton, 42-43, 130-131, 134-135, 158-159, 166-167, 172-173
skin, 130-131, 152-153, 166-167, 168-169
slate, 44-45
smell, 138-139, 154-155
smoke, 106-107
snail, 170-171
snake, 166-167, 168-169
snapper, 160-161
snow, 14-15, 32-33, 62-63, 64-65, 92-93, 94-95, 106-107, 108-109, 112-113, 116-117, 118-119, 192-193, 240-241, 244-245.
soil, 84-85, 92-93
  characteristics, 84-85
    humus, 84-85
    subsoil, 84-85
    bedrock, 84-85
  formation, 84-85
  types, 84-85:
    desertic
    laterite
    permafrost
solfatara, 36-37
solstice, 16-17
sonar, 150-151
soredia, 208-209
sori, 212-213
Southern Alps (New Zealand), 68-69
sparrow, 152-153
species, 82-83, 88-89, 90-91
sperm whale, 146-147, 240-241
spider, 174-175
spiracle, 146-147, 150-151

sponge, 96-97
sporangiophore, 214-215
sporangium, 212-213, 214-215
spore, 208-209, 210-211, 212-213, 214-125, 216-217, 218-219, 220-221, 222-223
sporophyte, 210-211, 212-213
spring, 36-37
sprout, 190-191
stem, 190-191, 216-217
stigma, 194-195
stipe, 208-209
stoma, 192-193
storm, 110-111, 112-113, 114-115, 116-117, 118-119, 122-123, 124-125, 126-127.
strata (soil), 92-93
stratopause, 106-107
stratosphere, 24-25, 106-107, 110-111, 116-117
streambed, 30-31
sturgeon, 160-161
style, 194-195
subduction, 26-27, 28-29, 40-41, 50-55, 66-67, 68-69, 78-79
subphylum, 82-83
subsoil, 84-85
Sun, 14-15, 16-17, 18-19, 38-39, 98-99, 100-101, 102-103, 106-107, 206-207
    solar activity, 100-101
    solar energy, 38-39, 86-87, 106-107
    solar radiation, 14-15, 98-99, 102-103, 106-107
    solar wind, 18-19, 100-101
superclass, 82-83
sweet violet, 178-179
syconium, 198-199
symbiosis, 96-97, 208-209, 216-217

T
taiga, 90-91, 108-109
tail, 136-137, 152-153
talc, 48-49
Tambora, mount, 64-65
taste, 138-139, 154-155
teeth, 142-143, 146-147
temperature, 108-109
tendrils, 192-193

tepal, 194-195
Tertiary Period, 22-23
thallus, 202-203, 204-205, 208-209-212-213, 214-21
thorax, 152-153, 166-167
thunder, 122-123
tick, 174-175
tiger, 140-141
tissue, 190-191
Timbuktu, 108-109
toad, 164-165
tongue, 138-139, 166-167
toe, 136-137
topaz, 48-49, 50-51
topi, 92-93
tornado, 124-125
Torres del Paine National Park (Chile), 46-47
totipotency, 186-187
touch, 154-155
transpiration, 32-33
tree, 180-181
trench, 78-79, 94-95
trilobites, 28-29, 42-43
tropism, 186-187
tropopause, 106-107
troposphere, 24-25, 98-99, 106-107, 110-111, 116-117
trout, 158-159
trunk, 180-181
tsunami, 74-75, 76-77
tuareg, 226-227
tuber, 190-191
tundra, 90-91, 92-93, 108-109
turquoise, 50-51
turtle, 166-167
typhoon, 124-125

U
ulva, 206-207
unguligrade, 136-137
Uranus, 18-19

V
Venus, 14-15, 18-19
vertebra, 168-169
viper, 168-169
vision, 140-141, 154-155
volcano, 22-23, 58-59, 60-61, 62-63, 64-65, 66-67
    location around the world,

66-67
pyroclastic flow, 64-65
pyroclastic products, 62-63
super volcanoes, 22-23
types of volcanoes, 60-61:
    caldera volcano
    cinder cone
    lava dome
    shield volcano
    stratovolcano
volcanic caldera, 58-59
volcanic eruption, 62-63
    effects, 64-65
types of effusive eruption, 62-63:
    fissure
    hawaiian
    types of explosive eruption, 62-63:
    pelean
    strombolian
    vesuvian
    vulcanian
volva, 216-217
vortex, 124-125

W
walnut, 178-179
warning, 100-101
    Planet Warming, 100-101
water, 14-15, 32-33, 34-35, 36-37, 38-39, 40-41, 54-55, 74-75, 84-85, 86-87, 90-91, 92-93, 94-95, 96-97, 98-99, 106-107, 108-109, 116-117, 118-119, 120-121, 132-133, 146-147, 150-151, 158-159, 162-163, 164-165, 172-173, 178-179, 180-181, 182-183, 184-185, 186-187, 188-189, 190-191, 202-203, 206-207, 226-227, 232-233, 240-241
    availability, 32-33
    cycle, 14-15, 32-33, 94-95
    evaporation, 14-15, 32-33, 94-95
    condensation, 14-15, 32-33, 94-95
    precipitation, 14-15, 32-33, 94-95
    circulation (runoff), 32-33, 94-95, 106-107

density, 94-95
    freshwater, 32-33, 94-95
    salt water, 32-33, 94-95
    water vapor, 14-15, 24-25, 32-33, 94-95, 106-107, 108-109, 112-113, 114-115, 116-117, 118-119, 120-121, 126-127
wave, 26-27, 68-69, 70-71, 112-113
    Rossby waves, 112-113
    seismic waves, 68-69, 70-71
    Rayleigh waves, 70-71
weather, 14-15, 24-25, 106-107, 110-111, 112-113, 114-115, 116-117, 118-119, 120-121, 122-123, 126-127
weathering, 40-41
Wegener, Alfred, 26-27
whiskey, 218-219
whorl, 194-195
wind, 14-15, 18-19, 28-29, 32-33, 34-35, 38-39, 40-41, 90-91, 92-93, 94-95, 106-107, 108-109, 110-111, 112-113, 114-115, 116-117, 120-121, 124-125, 126-127, 156-157, 170-171, 178-179, 190-191, 192-193, 194-195, 226-117, 232-233, 240-241, 246-247
wine, 222-223
wing, 148-149, 152-153, 156-157
woodlouse, 172-173

X
xylem, 180-181, 190-191, 192-193

Y
yak, 230-231
yeast, 222-223
Yellowstone National Park, 36-37

Z
zebra, 92-93
zoospore, 204-205
zygote, 210-211, 212-213

PHOTO CREDITS
Infographics Sol90Images
Photography Corbis, ESA, Getty Images,
Graphic News, NASA, National Geographic,
Science Photo Library.